510 D83w 7084278
Driver, Rodney D. 1932-
Why math?

Undergraduate Texts in Mathematics

Undergraduate Texts in Mathematics

Apostol: Introduction to Analytic Number Theory.
1976. xii, 338 pages. 24 illus.

Armstrong: Basic Topology.
1983. xii, 260 pages. 132 illus.

Bak/Newman: Complex Analysis.
1982. x, 224 pages. 69 illus.

Banchoff/Wermer: Linear Algebra Through Geometry.
1983. x, 257 pages. 81 illus.

Childs: A Concrete Introduction to Higher Algebra.
1979. xiv, 338 pages. 8 illus.

Chung: Elementary Probability Theory with Stochastic Processes.
1975. xvi, 325 pages. 36 illus.

Croom: Basic Concepts of Algebraic Topology.
1978. x, 177 pages. 46 illus.

Curtis: Linear Algebra: An Introductory Approach.
1984. x, 337 pages. 37 illus.

Dixmier: General Topology.
1984. x, 140 pages. 13 illus.

Driver: Why Math?
1984. xiv, 234 pages. 87 illus.

Ebbinghaus/Flum/Thomas Mathematical Logic.
1984. xii, 216 pages. 1 illus.

Fischer: Intermediate Real Analysis.
1983. xiv, 770 pages. 100 illus.

Fleming: Functions of Several Variables. Second edition.
1977. xi, 411 pages. 96 illus.

Foulds: Optimization Techniques: An Introduction.
1981. xii, 502 pages. 72 illus.

Foulds: Combinatorial Optimization for Undergraduates.
1984. xii, 222 pages. 56 illus.

Franklin: Methods of Mathematical Economics. Linear and Nonlinear Programming. Fixed-Point Theorems.
1980. x, 297 pages. 38 illus.

Halmos: Finite-Dimensional Vector Spaces. Second edition.
1974. viii, 200 pages.

Halmos: Naive Set Theory.
1974, vii, 104 pages.

Iooss/Joseph: Elementary Stability and Bifurcation Theory.
1980. xv, 286 pages. 47 illus.

Jänich: Topology.
1984. ix, 180 pages (approx.). 180 illus.

Kemeny/Snell: Finite Markov Chains.
1976. ix, 224 pages. 11 illus.

Lang: Undergraduate Analysis.
1983. xiii, 545 pages. 52 illus.

Lax/Burstein/Lax: Calculus with Applications and Computing, Volume 1. Corrected Second Printing.
1984. xi, 513 pages. 170 illus.

LeCuyer: College Mathematics with A Programming Language.
1978. xii, 420 pages. 144 illus.

continued after Index

R. D. Driver

Why Math?

With 86 Illustrations

Springer-Verlag
New York Berlin Heidelberg Tokyo

R. D. Driver
Department of Mathematics
University of Rhode Island
Kingston, RI 02881
USA

AMS Subject Classification: 00A06, 00A05

Library of Congress Cataloging in Publication Data
Driver, Rodney D. (Rodney David) date
Why math?
(Undergraduate texts in mathematics)
Includes index.
1. Mathematics—1961– . I. Title. II. Series.
QA39.2.D75 1984 510 84-3128

Typeset by Asco Trade Typesetting Ltd., Hong Kong.
Printed and bound by R. R. Donnelley and Sons, Harrisonburg, Virginia.
Printed in the United States of America.

9 8 7 6 5 4 3 2 1

ISBN 0-387-90973-7 Springer-Verlag New York Berlin Heidelberg Tokyo
ISBN 3-540-90973-7 Springer-Verlag Berlin Heidelberg New York Tokyo

Preface

This text aims to show that mathematics is useful to virtually everyone. And it seeks to accomplish this by offering the reader plenty of practice in elementary mathematical computations motivated by real-world problems.

The prerequisite for this book is a little algebra and geometry—nothing more than entrance requirements at most colleges.

I hope that users—especially those who "don't like math"— will complete the course with greater confidence in their ability to solve practical problems (without seeking help from someone who is "good at math").

Here is a sampler of some of the problems to be encountered:

1. If a U.S. dollar were worth 1.15 Canadian dollars, what would a Canadian dollar be worth in U.S. money?
2. If the tax rates are reduced 5% one year and then 10% in each of the next 2 years (as they were between 1981 and 1984), what is the overall reduction for the 3 years?
3. An automobile cooling system contains 10 liters of a mixture of water and antifreeze which is 25% antifreeze. How much of this should be drained out and replaced with pure antifreeze so that the resulting 10 liters will be 40% antifreeze?
4. If you drive halfway at 30 mph and the rest of the distance at 50 mph, what is your average speed for the entire trip?
5. A tank storing solar heated water stands unmolested in a room having an approximately constant temperature of 80°F. If the tank cools from 120°F to 100°F in 3 days, what will be its temperature after 3 more days?
6. If a 10-inch pizza costs $3.00, what should a 15-inch pizza cost, assuming no discount for quantity?
7. How can one lay out a *rectangle* 24 by 32 feet for the foundation of a house?
8. Why does the horn of a locomotive (apparently) change its pitch as it passes you, and what does this have to do with radar speed traps?

9. How could Voyager 1 send back beautiful color pictures of Saturn one billion miles away using only a 20-watt transmitter?
10. Your office is on the ninth floor of a 12-story building, but it seems that whenever you want to go *down*, the elevator is going *up*. Is this just a curious (and annoying) coincidence?

The "obvious" answers to Problems 1 through 6, for most people, are 85 cents, 25%, 1.5 liters, 40 mph, 80°F, and $4.50. And each of these answers is *wrong*. I hope that by the time you finish this course, these "obvious" answers will be clearly wrong and the correct answers will be practically obvious.

Incidentally, if you already know how to correctly handle most of the above ten questions, you probably should not waste your time taking this course.

Other topics considered in examples and problems include:

computations of discounts, surcharges, interest, life insurance rates, and the required amount of gravel for a driveway or concrete for a foundation,

studies of the economics of small versus large packages, heat loss from small vs. large buildings, and the longevity of slow vs. fast space travelers,

explanations of the principles of levers, navigation, carbon dating, and tax-sheltering investment accounts,

appraisals of the hazards of smoking, gambling, and civil-defense planning.

The book was designed for the basic general mathematics courses required by most universities and colleges. But experience shows that science and engineering students can also benefit from it.

Notes to the Instructor. Despite the stated prerequisites of college entrance, I always find it essential to spend a week on the review material in Chapter 1.

I have never been able to cover more than two-thirds of this text in any one semester. So it is designed for flexible choice of topics. After Chapters 1 through 4, you can proceed to Chapter 5, 6, 7, 8, 10, or 11. (Chapter 9 relies on 8 and Chapters 12 and 13 require 11.)

Chapters 1 through 4 are intended to be tractable without a calculator, and I generally do not permit calculators during tests until later in the course.

Throughout the text, the occasional problem which might be a bit more challenging than the average is identified with an asterisk. In several chapters the last (and more difficult) section is a good candidate for omission if necessary.

Acknowledgments. Valuable suggestions which have been used in this book were made by Bruce K. Driver, David M. Driver, Edmund A. Lamagna, Lewis I. Pakula, Abe Shenitzer, and Robert C. Sine. I gratefully acknowledge their help.

I also want to thank the staff at Springer-Verlag New York for their support and cooperation.

The cartoons on pages 86 and 142 by Sidney Harris first appeared in *American Scientist*. The rest of the drawings were produced by J & R Services from my crude sketches.

My final thanks go to Carole for her continuing help and encouragement.

Contents

Applications Discussed xi

Chapter 1
Arithmetic Review 1

1.1 Basis Rules 1
1.2 Division, Fractions, and Exponents 6
1.3 Percentages 11
1.4 Rates 16

Chapter 2
Prime Numbers and Fractions 21

2.1 Prime Numbers and Factorization 21
2.2 Greatest Common Factor 26
2.3 Rationals and Irrationals 30

Chapter 3
The Pythagorean Theorem and Square Roots 33

3.1 The Theorem 33
3.2 Square Roots Which Are Irrational 37
3.3 Computation of Square Roots by Successive Approximation 40

Chapter 4
Elementary Equations 44

4.1 Equations in One Unknown 44

4.2 The Use of Two or More Unknowns 50
4.3 Graphing 54

Chapter 5
Quadratic Polynomials and Equations 62

5.1 Solution of Quadratic Equations 62
5.2 Applications of Quadratic Equations 67
5.3 Quadratic Polynomials 71

Chapter 6
Powers and Geometric Sequences 77

6.1 Applications of Powers 78
6.2 More on Half-Lives 81
6.3 Compound Interest and Related Matters 87
6.4 IRAs and Similar Tax Sheltered Accounts 91
6.5 Geometric Series—the "Sum" of a Geometric Sequence 96

Chapter 7
Areas and Volumes 103

7.1 Areas 103
7.2 Volumes 108
7.3 Surface Area of a Solid (versus Volume) 115
7.4 Computation of Cube Roots 118

Chapter 8
Galilean Relativity 123

8.1 Displacement and Velocity Vectors 123
8.2 Doppler Effect 128
8.3 Components of Vectors 135

Chapter 9
Special Relativity 143

9.1 Simultaneity and Einstein's Postulate 143
9.2 Time Dilation 148
9.3 Length Contraction 151

Chapter 10
Binary Arithmetic 156

10.1 Decimal, Binary, and Ternary Representation of Integers 156

10.2 Subtraction and Division in Base Two 160
10.3 Applications 164

Chapter 11
Sets and Counting 170

11.1 Set Notation 170
11.2 Counting 174

Chapter 12
Probability 178

12.1 Elementary Ideas and Examples 178
12.2 Mutually Exclusive Events 183
12.3 The Basic Rules 187
12.4 Quality Control (optional) 194
12.5 Expectation 200
12.6 Conditional Probability 205

Chapter 13
Cardinality 212

13.1 Countable Sets 212
13.2 Countably Many Countable Sets 215
13.3 The Reals vs. the Rationals 217

Answers to Odd-Numbered Problems 221

Index 229

Applications Discussed

[*Note*. The page listed may be followed by others on the subject.]

Animals and insects
  air resistance, 118
  size limits, 117
Archaeology
  carbon dating, 82
  dinosaur, 117
  Great pyramid, 37
Area
  acreage, 107
  container, 117
  house lot, 107
  lawns, 107
  maximum enclosed, 74
  pizzas, 106
Astronomy
  age of universe, 147
  distances, 134, 147
  red shift, 134
  Voyager, 134, 147, 167
Automobile
  antifreeze, 46
  breakdowns, 193
  fatalities, 211
  gas mileage, 20
  speed and distance, 45

Binary numbers
  bit, 166
  byte, 166
  data transmission, 165

Civil Defense,
  crisis relocation, 193
  fallout protection, 114
Clocks and relativity
  clock rates, 149
  clock synchronization, 144
  time dilation, 148
  second, 148
Communication
  binary signals, 166
  data encryption, 29
  Morse code, 165
  radio, 164
Construction
  driveway gravel, 113

Construction (*cont.*)
foundations, 43, 114
lumber sizes, 42, 113
square corners, 35
survey distances, 43
Cooking
boiling eggs, 114
cooling, 79, 84
recipe adjustment, 12
Costs and economies
container sizes, 11, 117
coupons, 11
discounts, 13
electricity, 16, 87
heating, 87, 118
hose sizes, 107, 117
mileage, 20
pizza sizes, 106
sales charges, 14

Doppler effect
falling bomb, 141
red shift, 134
sound, 129, 138
speed detectors, 132, 141

Energy
conservation, 19, 87
heat losses, 79, 86, 118
solar heat storage, 80, 86, 117

Financial Matters
compound interest, 87, 102
currency exchange, 45
discounts, 13
down payments, 12, 48
income tax, 15, 60
inflation rates, 90
interest and dividends, 15, 51
IRAs, 91
investment fund charges, 14, 91
life insurance, 201
money market statements, 19
national debt, 11
pyramid sales schemes, 80
raises, 20
sales tax, 12
shopping coupons, 11
tax sheltering, 91
zero coupon CDs, 89, 120
Free fall
ball bouncing, 99
bomb, 141
sky rocket, 70

Gambling
expectation, 81, 202
fair games, 203
lottery, 205
Monte Carlo fallacy, 193, 206
odds, 203
poker hands, 175
probabilities, 183, 187
ruin, 204
Gardening
fertilizer, 52, 107, 114
flower pots, 114, 120
hoses, 107, 117
lawns, 107, 114
seeds, 190, 201
Gold
coin sizes, 113
Hieron's crown, 112

Health and smoking
heart disease, 211
lung cancer, 185, 208

Levers
lifting, 48
scales, 49
see saw, 49
Light
Doppler effect, 134

frequency, (color) 134
speed of, 132, 145
wavelengths, 134

Materials
carpet, 107
container, 117
display, 118
drapes, 107
lawn, 107, 114
paint, 108
yarn, 107, 113, 121
Metric units
area, 107
length, 11, 107
temperature—Celsius, 60
volume, 19, 113
weight, 114
Mixture concentrations
antifreeze, 46
fertilizer, 52
gas–oil, 49
medicine, 54

Navigation, across current
airplane, 127, 141
boat, 124, 137

Population
crowding, 118
growth, 89
Probability, other than gambling
automobile breakdowns, 193
elevator arrival, 182
fireflies, 211
life insurance, 201
quality control, 192, 194
seed germination, 190, 201
sex of children, 182, 193, 206
smoking and health, 185, 208
Pyramiding
chain letter, 78, 99
sales schemes, 80, 102

Radio
frequencies, 134
radar, 132
speed, 132
transmission, 164
wavelengths, 134
Radioactive materials
carbon dating, 82
half lives, 78
nuclear wastes, 86
Relativity
Galilean, 123
length contraction, 151
special, 143
time dilation, 148

Shopping
coupons, 11
discounts, 13
package sizes, 11, 111, 117
pizzas, 106
sales tax, 12, 48
Sizes
animals, 117
ball of string, 111, 121
buildings, 118
bushel, 121
coins, 112, 116, 122
cans, 111, 117
flower pots, 114, 120
foundation, 114
Hieron's crown, 112
hot tubs, 114
irregular shapes, 112
liter, 19
lumber, 42, 113
pizzas, 106
solar heat storage, 117
Sound
distance estimates, 49, 134
Doppler effect, 128

Sound (*cont.*)
frequency (pitch), 129
speed of, 20, 49
wavelength, 128
Space vehicles
communication, 134
relativity, 149, 153
Voyager, 134, 147, 167
Speed
airplane, 54, 70, 127, 141
average, 45
boat, 54, 67, 124, 137
light, 132
radar, traps 132
sound, 20, 49

Tax
income, 15, 60
sales, 12, 48
sheltering, 91

Warfare
civil defense, 114, 193
falling bomb, 141
plutonium, 86
Water
flow in hose, 117
leak, 19
Weight
boulder, 114
coin, 113
dirt, 114
metric, 114

CHAPTER 1

Arithmetic Review

Hardly any "practical" mathematics can be done at an elementary level unless one is comfortable with arithmetic.

This is not a course on arithmetic as such. So it is assumed that the reader is already quite familiar with that subject, perhaps needing just a little review. The brief review offered here puts particular emphasis on the distributive law, on manipulation of fractions, and on percentages.

However, even this basic chapter also covers some important applications. The section on percentages includes examples and problems on sales taxes, discounts, recipe adjustments, and comparison of investment programs. The section on rates examines costs of electricity, interest earned on savings, and speed-time-distance problems.

1.1. Basic Rules

You undoubtedly know the **commutative laws** of addition and multiplication,

$$a + b = b + a \qquad \text{and} \qquad a \times b = b \times a,$$

where a and b are any two numbers. Most of us use these laws almost daily without giving them a thought.

For example, if you wanted the value of $3 + 39$, you could start with 3 and add 1 thirty nine times:

$$\begin{aligned} 3 + 1 &= 4, \\ 4 + 1 &= 5, \\ 5 + 1 &= 6, \\ &\vdots \\ 41 + 1 &= 42. \end{aligned}$$

But, instead, it is much easier to use the fact that $3 + 39 = 39 + 3$ and think

$$39 + 1 = 40,$$

$$40 + 1 = 41,$$

$$41 + 1 = 42.$$

Indeed, this is so trivial that one just reasons mentally

$$3 + 39 = 39 + 3 = 42.$$

Similarly, it would be a considerable chore to compute 2×13 interpreted as adding 2 to itself 13 times. Instead, one uses the commutative law of multiplication which gives $2 \times 13 = 13 \times 2$. Thus the result of adding thirteen 2s is the same as adding two 13s,

$$2 + 2 + 2 + 2 + 2 + 2 + 2 + 2 + 2 + 2 + 2 + 2 + 2 = 13 + 13 = 26.$$

As another example, note that the computation

$$\begin{array}{r} 3 \\ \times 139 \\ \hline 27 \\ 9 \\ 3 \\ \hline 417 \end{array}$$

is more cumbersome than

$$\begin{array}{r} 139 \\ \times \quad 3 \\ \hline 417 \end{array}.$$

But both are equivalent since $3 \times 139 = 139 \times 3$.

In practice, if asked to multiply 3 by 139, one automatically does the simpler calculation 139×3 instead, without stopping to worry about the justification.

A word about the various notations for multiplication is in order. The product of two numbers a and b can be written as

$$a \times b \quad \text{or} \quad a \cdot b \quad \text{or} \quad ab \quad \text{or} \quad a(b) \quad \text{or} \quad (a)b.$$

In general, use whichever notation seems more convenient. However, you should avoid the cross $\times$ whenever it might be mistaken for the letter x. When using the dot for multiplication, be careful to keep it high enough so that it is not mistaken for a decimal point. Thus $2 \cdot 3 = 2 \times 3$. When using letters to represent numbers, say a and b, the product is most compactly written as ab omitting the dot. But the product of two numerals always requires a dot or other symbol. Clearly $2 \cdot 3$ and $2(\frac{2}{3})$ *cannot* be written as 23 and $2\frac{2}{3}$.

The next well-known laws of arithmetic are the **associative laws**

$$(a + b) + c = a + (b + c) \qquad \text{and} \qquad a(bc) = (ab)c,$$

where a, b, and c are any numbers.

For example, if you wanted to evaluate $(29 + 127) + 3$, it would be best to think

$$(29 + 127) + 3 = 29 + (127 + 3) = 29 + 130 = 159.$$

You also should not waste time computing $(127 \times 5) \times 2$ directly. It is much easier to think

$$(127 \times 5) \times 2 = 127 \times (5 \times 2) = 127 \times 10 = 1270.$$

Get in the habit of looking for shortcuts!

The law of arithmetic which causes trouble for *most* people is the **distributive law**

$$a(b + c) = ab + ac.$$

So this deserves some special emphasis.

Note first that, by virtue of the commutative law of multiplication, an equivalent statement of the distributive law is

$$(b + c)a = ba + ca.$$

Make sure that you understand the following examples.

Example 1

$$2(3 + 4) = 2 \cdot 3 + 2 \cdot 4$$
$$(a + b)7 = 7(a + b) = 7a + 7b$$

One must be careful to use parentheses when they are needed.

Example 2. With parentheses,

$$(6 + 4)\tfrac{1}{2} = 5.$$

But if you accidentally forgot the parentheses you would have

$$6 + 4 \cdot \tfrac{1}{2} = 8,$$

quite a different value.

Example 3. A number of important applications later in this chapter, and elsewhere in the text, will use the fact that

$$a + ba = (1 + b)a.$$

When this first appears, someone always asks, "Where did the 1 come from?" It is simply a matter of recognizing that $a = 1a$, and then applying the distributive law:

$$a + ba = 1a + ba = (1 + b)a.$$

A specific numerical example is

$$9 + (0.2)9 = (1 + 0.2)9 = (1.2)9.$$

Confirm this by computing separately $9 + (0.2)9$ and 1.2×9. You should get the same answer each way.

Now consider subtraction, and the related concept of negative numbers. The difference $b - a$ (read "b minus a" or "a subtracted from b") is defined as that number which when added to a gives b. That is,

$$b - a = c \qquad \text{if} \qquad a + c = b.$$

For example, since $2 + 3 = 5$, we agree that $5 - 2 = 3$. In general, there is no problem with the symbol $b - a$ when $b > a$ (b is greater than a).

But what is the meaning of $b - a$ when $a > b$? For example, what is the meaning of $2 - 5$ or $0 - 3$?

The number -3 (read "negative three") was invented to satisfy the equation

$$3 + c = 0.$$

Thus we say $0 - 3 = -3$. Furthermore, invoking the associative law of addition,

$$5 + (-3) = (2 + 3) + (-3) = 2 + [3 + (-3)] = 2 + 0 = 2.$$

Thus we conclude that $2 - 5 = -3$.

In general, $-a$ is defined to satisfy the equation

$$a + (-a) = 0.$$

The following rules apply, regardless of whether a and b are positive or negative numbers:

$$b - a = -(a - b),$$

$$-a = (-1)a,$$

$$-(-a) = a,$$

$$(-a)(-b) = ab,$$

and

$$b - a = b + (-a).$$

The last of these says that subtracting a is the same as adding $-a$ (and this is true whether a is positive or negative).

Thus, for example,

$$17 - (-5) = 17 + 5 = 22.$$

Confirm the assertions of the following.

Example 4

$$27 + (-15) = 27 - 15 = 12,$$

$$12 - (-3) = 12 + 3 = 15,$$

$$-12 - (-3) = -12 + 3 = -9,$$

$$(-3)(-2) = 6,$$
$$-(-5) = 5,$$
$$6 - 27 = -(27 - 6) = -21.$$

The next three examples will be used in applications later. Each involves repeated use of the distributive law.

Example 5

$$\begin{aligned}(a + b)(c + d) &= a(c + d) + b(c + d)\\ &= ac + ad + bc + bd.\end{aligned}$$

Example 6. Recall that a^2 stands for $a \cdot a$. Then compute

$$\begin{aligned}(a + b)^2 &= (a + b)(a + b)\\ &= a(a + b) + b(a + b)\\ &= a^2 + ab + ba + b^2\\ &= a^2 + 2ab + b^2.\end{aligned}$$

Example 7 ("difference of two squares")

$$\begin{aligned}(a + b)(a - b) &= a(a - b) + b(a - b)\\ &= a^2 - ab + ba - b^2\\ &= a^2 - b^2.\end{aligned}$$

Putting $b = 1$ in Example 7 gives a useful trick for computing the product of two consecutive odd integers or two consecutive even integers. For instance,

$$19 \cdot 21 = (20 - 1)(20 + 1) = 20^2 - 1^2 = 399.$$

Make sure that you thoroughly understand each step of the computations in Examples 5, 6, and 7. If the results are not already familiar, you should memorize those of Examples 6 and 7 for future use.

PROBLEMS

1. Simplify each of the following expressions:
   (a) $2 \cdot (5 + 2)$ (b) $2 \cdot 5 + 2$ (c) $3(1 + a)$
   (d) $5[2 + 3(4 - 1)]$ (e) $2(-13)(-5)$ (f) $0.17 \cdot 10$
   (g) $100(1 + 0.06)$ (h) $(0.2)(0.03)$ (i) $13 - (-26)$
   (j) $-9 - (-26)$ (k) $(a - b)^2$ (l) $(2x + 1)(x - 7)$

2. Simplify:
   (a) $16 - 31$ (b) $a - (-b)$ (c) $16 - (-31)$
   (d) $(-2)(5 - 2)$ (e) $(-2)(a - b)$ (f) $5 - 0.17$
   (g) $0.5 - 17$ (h) $0.5 - 1.7$ (i) $0.1(1 - 2.5)$

3. Which of the following statements are incorrect?
   (a) $(1 + a)b = 1 + ab$ (b) $a(x - y) = ax - ay$
   (c) $(-a)(b - c) = -ab - ac$ (d) $9 - 6 \cdot \frac{1}{3} = 1$
   (e) $(a + b)^2 = a^2 + b^2$ (f) $a^2 - b^2 = (a - b)^2$
4. Why is $(1 + 0.1)^2$ *approximately* equal to $1 + 0.2$? See Example 6.
5. Rewrite each of the following as a single product without performing any multiplication (as in Example 3):
   (a) $10 + (0.05)(10)$ (b) $39 + 39(0.15)$ (c) $27 - (0.1)27$
6. Compute each of the following in your head (and quickly):
   (a) $(25 \cdot 17)4$ (b) $37 \cdot 7 + 7 \cdot 13$ (c) 1.3×10^2
   (d) $289 + 518 + 11$ (e) $14 \cdot 16$ (f) $29 \cdot 31$
7. Why is $1 - (0.99)^2$ *approximately* equal to 0.02? See Example 7.
8. Multiply out (a) $(x - 2y)(x + y)$ and (b) $(x - 2y)(x + 2y)$.

1.2. Division, Fractions, and Exponents

Just as negative numbers were invented to make sense out of subtraction, so fractions were invented to meet the needs of the division process. Let us, therefore, begin by reviewing the meaning of division.

Recall that division of a by b ($\neq 0$) can be denoted by any of three notations,

$$a \div b \qquad \text{or} \qquad \frac{a}{b} \qquad \text{or} \qquad a/b.$$

Division is defined as follows. If $b \neq 0$, then

$$a \div b = c \qquad \text{if} \qquad bc = a.$$

For example, $6 \div 3 = 2$ because $3 \cdot 2 = 6$.

But what is $2 \div 3$?

Since there is *no integer* c such that $3c = 2$, we *define* the answer to be the "fraction" $\frac{2}{3}$, also written as $2/3$.

Thus the symbol $\frac{2}{3}$ or $2/3$ may represent either the division process $2 \div 3$ or the number two-thirds. This "ambiguity" of the symbol $2/3$ does not matter since both interpretations, $2 \div 3$ and two-thirds, give the same value.

In general, an expression a/b where a and b are *both integers*, and $b \neq 0$, is called a **fraction**. For example,

$$1/2 \qquad 2/3, \qquad 4/3, \qquad \text{and} \qquad -2/5$$

are fractions. However, $2.7/1.3$ is not a fraction since the numbers 2.7 and 1.3 are not integers. Thus the interpretation of $2.7/1.3$ must be $2.7 \div 1.3$.

Regardless of whether a and b are integers, the result of the division a/b ($b \neq 0$) remains unchanged if both the numerator a and the denominator b are *multiplied* (*or divided*) *by the same nonzero number c*. That is,

$$\frac{a}{b} = \frac{ac}{bc} \quad \text{and} \quad \frac{a}{b} = \frac{a/c}{b/c} \quad \text{if } b \neq 0 \text{ and } c \neq 0.$$

A fraction a/b is said to be in *lowest terms* if there is no integer, other than 1, which is a factor of both the numerator and denominator.

Example 1. Reduce 34/51 to an equivalent fraction in lowest terms.

Solution

$$\frac{34}{51} = \frac{2 \times 17}{3 \times 17} = \frac{2}{3}.$$

The fraction $\frac{2}{3}$ is in lowest terms since there is no common integer factor of the numerator and denominator other than 1.

Example 2. The expression 2.7/1.3 is not a fraction since 2.7 and 1.3 are not integers. However, it is easy to find an equivalent expression which is a fraction:

$$\frac{2.7}{1.3} = \frac{2.7 \times 10}{1.3 \times 10} = \frac{27}{13}.$$

Example 3. Convert 0.12 into an equivalent fraction in lowest terms.

Solution

$$0.12 = \frac{12}{100} = \frac{3 \cdot 4}{25 \cdot 4} = \frac{3}{25}.$$

To determine which of two given fractions is the larger, one should convert them to two (equivalent) fractions with a *common* denominator.

Example 4. Determine which is larger, 5/9 or 4/7.

Solution. The two given fractions are equivalent to a pair of other fractions having the common denominator $9 \times 7 = 63$. Indeed,

$$\frac{5}{9} = \frac{5 \cdot 7}{9 \cdot 7} = \frac{35}{63} \quad \text{and} \quad \frac{4}{7} = \frac{4 \cdot 9}{7 \cdot 9} = \frac{36}{63}.$$

Since $36/63 > 35/63$, it follows that $4/7 > 5/9$.

If a/b and c/d are fractions, their *product* is obtained by multiplying together their numerators and denominators, respectively. Thus

$$\frac{a}{b} \cdot \frac{c}{d} = \frac{ac}{bd}.$$

Example 5

$$\frac{2}{3}\cdot\frac{6}{7}=\frac{2\cdot 6}{3\cdot 7}=\frac{2\cdot 2}{7}=\frac{4}{7}.$$

Division of a/b by c/d is accomplished by inverting the divisor, c/d, and multiplying. Thus

$$\frac{a}{b}\div\frac{c}{d}=\frac{a}{b}\cdot\frac{d}{c}=\frac{ad}{bc}.$$

Example 6

$$\frac{2}{3}\div\frac{6}{7}=\frac{2}{3}\cdot\frac{7}{6}=\frac{2\cdot 7}{3\cdot 6}=\frac{7}{9}.$$

In order to *add* (or *subtract*) two fractions, one must first convert them to equivalent fractions with a common denominator. How can one compute

$$\frac{a}{b}+\frac{c}{d}?$$

Convert the two fractions, a/b and c/d, into equivalent fractions having the denominator bd. Thus,

$$\frac{a}{b}+\frac{c}{d}=\frac{ad}{bd}+\frac{bc}{bd}=\frac{ad+bc}{bd}.$$

In numerical problems one may find that b and d have some common factor(s). Then it is better to work with the least common denominator, rather than the larger number bd.

Example 7. Add $\frac{1}{4}$ and $\frac{5}{6}$.

Solution. Converting to two equivalent fractions with the least common denominator, 12, you find

$$\frac{1}{4}+\frac{5}{6}=\frac{3}{12}+\frac{10}{12}=\frac{13}{12}.$$

Note that if a fraction involves a negative sign, it does not matter whether that minus sign is put in the numerator or the denominator or in front of the fraction. In other words,

$$\frac{-a}{b}=\frac{a}{-b}=-\frac{a}{b}.$$

Example 8. Simplify $\frac{1}{2}(\frac{3}{4}-\frac{4}{5})$.

Solution

$$\frac{1}{2}\left(\frac{3}{4}-\frac{4}{5}\right)=\frac{1}{2}\left(\frac{15}{20}-\frac{16}{20}\right)=-\frac{1}{2}\cdot\frac{1}{20}=-\frac{1}{40}.$$

This section concludes with a word about *exponents*.

Positive integer exponents 1, 2, 3, 4, ... have the following meanings:

$$a^1 = a,$$
$$a^2 = a \cdot a,$$
$$a^3 = a \cdot a \cdot a,$$
$$a^4 = a \cdot a \cdot a \cdot a,$$

and so on. When defined this way, a^n is called the nth *power* of a.

Using this notation, consider an extension of Example 6 of Section 1.1.

Example 9

$$\begin{aligned}(a+b)^3 &= (a+b)(a+b)^2\\ &= (a+b)(a^2+2ab+b^2) \qquad \text{(calculated previously)}\\ &= a(a^2+2ab+b^2)+b(a^2+2ab+b^2)\\ &= a^3+2a^2b+ab^2+ba^2+2bab+b^3\\ &= a^3+3a^2b+3ab^2+b^3.\end{aligned}$$

Zero and negative exponents are defined, when $a \neq 0$, by

$$a^0 = 1,$$
$$a^{-1} = 1/a,$$
$$a^{-2} = 1/a^2,$$
$$a^{-3} = 1/a^3,$$

etc.

To recall the laws (or theorems) for manipulation of integer exponents, review the following:

Example 10

$$a^3a^2 = (a \cdot a \cdot a)(a \cdot a) = a^5,$$

$$a^3a^{-2} = (a \cdot a \cdot a) \cdot \frac{1}{a \cdot a} = a = a^{3+(-2)} \qquad (\text{if } a \neq 0),$$

$$\frac{a^3}{a^2} = \frac{a \cdot a \cdot a}{a \cdot a} = a = a^{3-2} \qquad (\text{if } a \neq 0),$$

$$\frac{a^2}{a^3} = \frac{a \cdot a}{a \cdot a \cdot a} = \frac{1}{a} = a^{-1} = a^{2-3} \qquad (\text{if } a \neq 0),$$

$$\frac{a^3}{a^{-2}} = \frac{a \cdot a \cdot a}{1/(a \cdot a)} = (a \cdot a \cdot a)(a \cdot a) = a^5 = a^{3-(-2)} \quad (\text{if } a \neq 0),$$

$$(a^2)^3 = (a \cdot a)^3 = (a \cdot a)(a \cdot a)(a \cdot a) = a^6 = a^{2 \cdot 3},$$

$$(ab)^3 = (ab)(ab)(ab) = a^3b^3,$$

$$\left(\frac{a}{b}\right)^4 = \frac{a}{b} \cdot \frac{a}{b} \cdot \frac{a}{b} \cdot \frac{a}{b} = \frac{a \cdot a \cdot a \cdot a}{b \cdot b \cdot b \cdot b} = \frac{a^4}{b^4} \qquad (\text{if } b \neq 0).$$

The eight calculations in Example 10 are illustrations of the following five laws for manipulation of integer exponents:

If n and m are any two integers, and if $a \neq 0$ and $b \neq 0$, then

$$a^n a^m = a^{n+m} \qquad \text{(first and second in Example 10)},$$

$$a^n/a^m = a^{n-m} \qquad \text{(third, fourth, fifth in Example 10)},$$

$$(a^n)^m = a^{nm} \qquad \text{(sixth in Example 10)},$$

$$(ab)^n = a^n b^n \qquad \text{(seventh in Example 10)},$$

$$(a/b)^n = a^n/b^n \qquad \text{(eighth in Example 10)}.$$

Example 11. As an application, notice how simple it becomes to compute $12^4 \div 4^3$:

$$\frac{12^4}{4^3} = \frac{3^4 \cdot 4^4}{4^3} = 3^4 \cdot 4 = 324.$$

Problems

1. Reduce each of the following to a fraction in lowest terms.
   (a) 2/6 (b) 9/111 (c) 85/50 (d) 1.4/0.3 (e) 0.016

2. Simplify each of the following, ending up with a fraction in lowest terms (or an integer).
   (a) $\frac{2}{5} \cdot \frac{3}{4}$ (b) $\frac{2}{5} \div \frac{3}{4}$ (c) $\frac{2}{5} + \frac{3}{4}$
   (d) $\frac{2}{5} - \frac{3}{4}$ (e) $\frac{4}{3}\left(\frac{5}{6} - \frac{7}{3}\right)$ (f) $\frac{1}{300} + \frac{1}{110}$

3. Simplify each of the following.
   (a) $\left(\frac{25}{6} \cdot \frac{15}{40}\right) \div \frac{5}{8}$ (b) $\frac{1}{1/2}$ (c) $\frac{1/2}{3 + 1/3}$
   (d) $\frac{5/4}{1 - 3/4}$ (e) $\frac{3/4 - 1/6}{1/2 + 2/3}$ (f) $\left(2 + \frac{5}{2}\right)\frac{1}{2}$

4. What is (a) one-eighth of 2/5? (b) $\frac{9}{2} \div 2$?

5. Determine the larger of each of the following pairs:
   (a) $\frac{3}{8}$ and $\frac{4}{9}$ (b) $\frac{7}{8}$ and $\frac{11}{13}$ (c) $\frac{15}{19}$ and $\frac{3}{4}$

6. Multiply out (a) $(27x)\frac{1}{x}$ and (b) $\left(x+\frac{5}{2}\right)^2$.

7. Which of the following statements are incorrect?
   (a) $\frac{1}{2}\left(2+\frac{1}{2}\right)=1+\frac{1}{2}$ (b) $\frac{1}{a}+\frac{1}{b}=\frac{1}{a+b}$ (c) $\frac{1}{a}+\frac{1}{b}=\frac{2}{a+b}$
   (d) $\frac{1}{a}+\frac{1}{b}=\frac{a+b}{ab}$

8. You are in a grocery store comparing a 20-ounce package of cereal for \$1.29 with a 25-ounce package of the same cereal for \$1.59. Which is cheaper per ounce?

9. Repeat Problem 8 assuming you have a 7-cent coupon good on any size package of this same cereal.

10. Repeat Problem 8 assuming you have a 20-cent coupon good on any size package of this same cereal.

11. Is there any reason why you might decide to use your coupon(s) on a larger size package even though the coupon makes the smaller package a better buy?

12. What is wrong with the following "proof"? Let $x = 1$. Then $x^2 - x = 2x - 2$. Thus $(x-1)x = (x-1)2$. $\therefore x = 2$. $\therefore 1 = 2$.

13. Simplify and express your answers without using exponents.
   (a) $2^4 \cdot 2^3$ (b) $3^5/3$ (c) $2^{-3}/2^2$
   (d) $10^{10} \cdot 10^{-8}$ (e) $2^{-2}/2^{-3}$ (f) $6^8/3^8$ (the easy way)

14. Do the following calculations quickly in your head:
   (a) $\frac{27}{16}\cdot\frac{4}{3}$ (b) 0.00013×10^6 (c) $\frac{13}{2} \div \frac{13}{4}$

15. Which of the following statements are incorrect?
   (a) $a^5 - a^3 = a^2$ (b) $3^4/3^{-2} = 3^2$ (c) $15^3/5^2 = \frac{15}{5} = 3$

16. (a) What is your "share" of the national debt? (b) What is your family's share? [*Hint*. For (a) compute (national debt) ÷ (U.S. population).]

17. A meter is approximately 39.37 inches. How many centimeters (hundreths of a meter) are in an inch?

1.3. Percentages

The expression ***p* percent of *a*** means p hundredths of a, or

$$\frac{p}{100}a.$$

The symbol % is often used in place of the word percent.

The following examples illustrate some typical situations involving percentages.

Example 1. If you were buying a house for \$38,000 and had to make a down payment of 25%, what would be the down payment?

Solution

$$\$38{,}000 \times \frac{25}{100} = \$38{,}000 \times 0.25 = \$9500.$$

Example 2. If the sales tax is 6% of a \$15.00 purchase, how much is the tax?

Solution

$$\$15 \times \frac{6}{100} = \$15 \times 0.06 = \$0.9 \quad \text{or} \quad 90 \text{ cents.}$$

Example 3. You are following a recipe which you want to increase by 50%. How much sugar should you use if the original recipe called for $\frac{2}{3}$ cup?

Solution

$$\frac{2}{3} + \frac{2}{3} \cdot \frac{50}{100} = \frac{2}{3}(1 + 0.5) = \frac{2}{3} \cdot \frac{3}{2} = 1 \text{ cup.}$$

Note the use of the distributive law here as in Example 3 of Section 1.1.

Example 4. Referring to Example 2, your total cost for an item priced at \$15.00 plus a 6% sales tax would be \$15.00 + \$0.90 = \$15.90. But there is a more efficient way of finding the total cost. Instead of computing the sales tax separately and adding it to the price, note that what you are really doing is increasing the price by 6%. Thus the total cost will be

$$15 + 0.06 \times 15 = (1 + 0.06)15 = (1.06)15 = 15.90.$$

Examples 3 and 4 are worth understanding. They are important applications of the distributive law. You should be able to arrive similarly at the following statements.

To increase a quantity by 50%, multiply it by 1.5. To increase a quantity by

6%, multiply it by 1.06. To increase by 17%, multiply the quantity by 1.17. Multiplication by 1.5 or 1.06 or 1.17 is easy on a calculator.

Example 5. A certain store gives a discount of 5% off the list price on cash purchases. If an item is listed at $15.00, what will its reduced "cash price" be at that store?

Solution. Now you must *reduce* the price by 5%. The result is

$$15 - \frac{5}{100} \cdot 15 = 15 - (0.05)15 = (1 - 0.05)15 = (0.95)15 = 14.25.$$

Replacing the 15 in this calculation by any other number, you should see that in order to reduce any quantity by 5% you need only multiply the quantity by 0.95. Similarly, to reduce a quantity by 17% multiply the quantity by 0.83, that is, $1 - 0.17$.

Example 6. A certain store gives a discount of 5% off the list price on cash purchases. However, it must also charge 6% sales tax. Some salespersons first deduct the 5% discount, and then add the 6% tax on the reduced (net) price. Others first compute and add the tax to the list price, and then compute 5% discount on the total price including tax. Which method results in the lower final cost?

Solution. Let the list price of an item be a. If the 5% discount is deducted first, the net price is

$$0.95a.$$

Now add the 6% sales tax to this to get a final cost of

$$(1.06)0.95a.$$

On the other hand, if one begins by adding the sales tax to the list price, one gets

$$1.06a.$$

Then deducting 5% of this total gives a final cost of

$$(0.95)1.06a.$$

Since $(1.06)0.95 = (0.95)1.06$, both methods have produced the *same* result.

Example 7. What *is* the final net effect of a 5% discount and a 6% sales tax? (It is *not* a 1% increase as might appear at first. Consider an article with a list price of $100. Then the discounted price is $95, so the 6% sales tax will be *less than* $6. And the net effect will be to increase the list price by slightly less than 1%.)

Solution. If the list price is a, the final cost of $(1.06)0.95a$ can be computed as follows:

$$\begin{aligned}(1.06)0.95a &= (1 + 0.06)(1 - 0.05)a\\ &= [1 + 0.06 - 0.05 - (0.06)(0.05)]a\\ &= [1 + 0.06 - 0.05 - 0.003]a\\ &= (1 + 0.007)a.\end{aligned}$$

Since $0.007 = 0.7/100$, the net effect is to increase the price by 0.7%, slightly less than 1% as predicted.

Example 8. Over a 5-year period the moneys entrusted to Investment Company A have grown by 60%. However, any money paid into the company was first reduced by a sales charge of 5%. (For example, if you had deposited $100 at the outset, your account would have *started* at $95.) No further charge is deducted when the money is withdrawn.

In the same time period the deposits in Investment Company B have also grown by 60%. Company B makes no initial sales charge. But when the money is withdrawn at the end of the period Company B deducts 5% of the (larger) payback amount.

The salesperson for Company A says that his or her company is better because the sales charge was only 5% of the initial investment, while the competition would have deducted 5% of the larger final amount. But the salesperson for Company B declares that it is better to make your entire initial deposit earn interest, and pay the charges later. Which one will you believe? (Someday, when you look into mutual funds or retirement accounts, you will actually meet these two sales representatives.)

Solution. If the intitial amount deposited was a, then with Company A it starts out as $0.95a$ and grows to

$$(1.6)0.95a.$$

The same amount deposited with Company B would grow to $1.6a$. But upon withdrawal this would be reduced by 5% to

$$(0.95)1.6a.$$

The outcomes are the same!

Notice how similar this was to Example 6.

Sometimes you will want to convert a fraction of some quantity into a percentage. This is illustrated in the next two examples.

Example 9. Three-fourths of something is what percentage?

Solution. By division, one finds that the fraction $\frac{3}{4} = 0.75$. So three-fourths of quantity a is

$$\frac{3}{4}a = 0.75a = \frac{75}{100}a.$$

It follows that $\frac{3}{4}$ is 75%.

Example 10. If a savings account balance was \$80.00 on July 1, and 1 year later it was \$85.00, by what percentage had it *increased*?

Solution. The increase was $\$85 - \$80 = \$5$. So the fractional increase was

$$\frac{\text{increase}}{\text{initial amount}} = \frac{5}{80} = \frac{1}{16}.$$

Now divide, as in the previous example, to find $\frac{1}{16} = 0.0625$. Since $0.0625 = 6.25/100$, the balance had increased by 6.25% (or $6\frac{1}{4}\%$).

Refer back to these examples, if necessary, as you work the problems.

PROBLEMS

1. How much cash do you need to make a 20% down payment on a car costing \$6800?
2. Find the *total* purchase price when a 7% sales tax is added to a list price of \$8.00.
3. You want to convert a list price into the total selling price including sales tax. By what single number should you multiply the list price if the sales tax is (a) 7%? (b) 8%? (c) $5\frac{1}{2}\%$?
4. You want to convert a list price into a net (discounted) price, assuming no sales tax. By what single number should you multiply the list price if the discount is (a) 2%? (b) 10%? (c) 15%?
5. Now you want to convert a list price into a final cost, taking into account both a discount and a sales tax. By what single number should you multiply the list price, and what final percentage increase or decrease does this represent if (a) the discount is 5% and the sales tax is 8%? (b) the discount is 10% and the sales tax is 7%?
6. You are increasing a recipe by 50%. How much soy sauce should you use if the original recipe called for $\frac{1}{4}$ cup?
7. Three-halves of something is what percentage?
8. Is a discount of 15% of the selling price of an article more than, less than, or the same as a discount of 10% of the selling price plus a discount of 5% of the reduced price?
9. If the federal income tax rates were cut by 5% one year and then 10% in each of the next 2 years, what would be the overall reduction in rates for the 3 years?
10. A store gives a discount of $d\%$ on a certain sale (where $0 < d < 100$) and adds a sales tax of $t\%$. Show why it does not matter to the customer whether the discount or the tax is computed first.
11. The shares of a certain investment fund cost \$10 each initially. At the end of a year each share was worth \$12. After a second year it was worth \$20, and after a third year \$16.50. Find the percentage change in value of a share (a) in the first year, (b) in the second year, and (c) in the third year.

12. For the investment fund described in Problem 11, the shares increased in value by 65% over the 3-year period. Do the answers to parts (a), (b), and (c) of Problem 11 add up to 65%? Explain.

13. Three-fourths exceeds five-eighths by what percentage?

14. This problem refers to Example 8. (a) If you had deposited $1000 with Investment Company A at the beginning of the 5-year period, how much could you withdraw at the end? (b) Answer the same question for Investment Company B.

15. If a retailer sets his selling price for an item by adding 50% to his (wholesale) cost, what percentage of his *selling price* is his (gross) profit?

16. Assume that the height of a TV screen is three-fourths of its width. Then its width exceeds its height by what percentage?

17. While away on a trip you buy a camera and some film priced at $57.96, and the cashier charges you $60.86. What is the sales tax (percentage) in that state?

18. Try to answer the following without using paper or pencil. Find the cost to the nearest dollar of
    (a) an overcoat discounted 40% from the list price of $69.99,
    (b) a home computer discounted 15% from the list price of $2000,
    (c) a home computer discounted 15% from the list price of $2100 (making use of your result in part (b)),
    (d) a black-and-white TV set with a 20% discount from the list price of $84.95.

1.4. Rates

If (in some endeavor) progress is being made at a certain rate, the total progress in a given time interval is the product of the rate and the time. Perhaps the most natural example is the speed of an automobile (the rate), and the distance covered in a given time (the total progress). Thus

$$\text{speed} \times \text{time} = \text{distance covered}.$$

Example 1. A motorist traveling at a speed of 30 mph for 4 hours will progress or travel a distance of

$$\text{speed} \times \text{time} = 30 \times 4 = 120 \text{ miles}.$$

Similarly, if some commodity is being consumed at a certain rate, then the total consumption in a given time interval is the product of the rate and the time.

Example 2. A 100-watt lamp left on for 24 hours uses

$$\text{rate} \times \text{time} = 100 \times 24 = 2400 \text{ watt-hours}$$

of electrical energy. The 100 watts represents a rate of consumption of energy; the 2400 "watt-hours" is the total amount of energy used.

The standard unit for measuring home usage of electrical energy is the kwh (kilowatt-hour) representing 1000 watt-hours. So the energy consumption in this example is commonly stated as 2.4 kwh.

These examples and others to follow can be unified in the format

$$\text{rate} \times \text{time} = \text{amount used (or gained)}. \tag{$*$}$$

Equation $(*)$ is simple. But to use it, you must recognize which information represents the rate and which represents a total amount. And you must make sure that the units are consistent.

Example 3. If a car is traveling at 30 mph, how far will it go in 7 seconds?

Solution. It would not make sense to multiply a speed in miles per hour by a time in seconds. One must either convert the 30 mph into miles-per-second or perhaps feet-per-second units, *or* convert the 7 seconds into a fraction of 1 hour. *Then* you can multiply.

In this example it is reasonable to convert the speed into feet-per-second units. Since there are 5280 feet in a mile and 3600 seconds in an hour,

$$\begin{aligned} 30 \text{ miles/hour} &= 30 \times 5280 \text{ feet/hour} \\ &= 30 \times \frac{5280}{3600} \text{ feet/second} \\ &= 44 \text{ feet/second.} \end{aligned}$$

So, in 7 seconds, the car will travel

$$\text{distance} = \text{speed} \times \text{time} = 44 \times 7 = 308 \text{ feet.}$$

Example 4. According to the Nielsen Television Index, a typical household watches television about 7 hours per day. How many kwh does this take per month assuming a color TV set using 300 watts?

Solution. The TV set consumes

$$300 \times 7 = 2100 \text{ watt-hours each day,}$$

or about

$$2100 \times 30.5 = 64{,}050 \text{ watt-hours each month.}$$

This is about 64 kwh per month.

Interest on savings can also be discussed using equation (*).

Example 5. If you leave $250 in a savings account earning interest at a rate of 5% a year, how much interest will be earned in 3 months?

Solution. The (annual) interest is 5% of the amount on deposit, namely $250 \times 5/100 = 12.50$ dollars per year. So, since 3 months is $\frac{1}{4}$ year, the interest earned in 3 months will be

$$\text{rate} \times \text{time} = 12.50 \times \tfrac{1}{4} = 3.125 \text{ dollars.}$$

In the next two examples you are not given the rate and the time. Instead, in one you are given the rate and the total, and asked to find the time; in the other you are given the time and the total, and asked to find the rate.

Example 6. How long would it take to travel 10 miles at a speed of 50 mph?

Solution. Here time is the unknown quantity. In order to use equation (*) it is convenient to invent a temporary symbol, say t, for the unknown time in hours. Then speed $\times$ time $=$ distance becomes

$$50 \times t = 10.$$

Divide both sides of this equation by 50 to get

$$t = \frac{10}{50} = \frac{1}{5} \text{ hour,}$$

i.e., 12 minutes.

Example 7. A statement from a money market fund shows that the balance increased from $2,440.78 to $2,460.73 in a 35-day period while no deposits or withdrawals were made. Determine the effective *annual* rate of interest earned.

Solution. The balance increased by $2{,}460.73 - 2{,}440.78 = 19.95$ dollars or by

$$\frac{19.95}{2{,}440.78} \times 100 = 0.817\%$$

in the 35 days. To interpret this as an annual rate, simply divide by 35 and multiply by 365. Thus the effective annual rate was

$$0.817 \times \frac{365}{35} = 8.52\%.$$

Actually, money market rates fluctuate from day to day. The above calculation yields a sort of average rate for the period.

Problems

1. How far would a car travel (a) in 40 minutes at 54 mph?, (b) in 2 seconds at 45 mph?

2. (a) This monthly electric bill shows a total consumption of 330 kwh and a charge of $30.75. What is the cost per kilowatt-hour?

Narragansett Electric

Month JAN84

4145 QUAKER LANE
N KINGSTOWN 02852
TEL 539-2367

Previous Bill PAY12727	26.04 26.04CR
Balance Due	.00

From	To	Rate	Previous Reading	Present Reading	KWH used	Description	Current Charges
DEC09	JAN11	A11	16119	16449	330	ELECTRIC	30.75

(b) Find the cost per kilowatt-hour from your own electric bill.

Electric rates can be very different in different localities. For problems below which require a cost for electrical energy, answers at the back of the book are based on an arbitrary rate of 8 cents per kwh.

3. Assume that your 1000-watt toaster toasts two slices of bread in 2 minutes. (a) How many kwh does this use? (b) What does it cost?

4. You return home from a 10-day trip and discover that you unintentionally left two 100-watt lights on. What will this mistake cost?

5. Assume that when no hot water is being used, your electric hot water heater uses energy at an average rate of 200 watts just to maintain the water temperature. How much could you save per month by keeping the water heater turned off except for 1 hour per day? [*Note.* An average month has 730 hours, i.e., $24 \times 365 \div 12$.]

6. Determine the monthly cost of the electrical energy for
   (a) three electric clocks, each rated at 2 watts,
   (b) a dehumidifier using energy at a constant rate of 250 watts,
   (c) a refrigerator-freezer which rests half the time, and which uses energy at an average rate of 350 watts when it is running.

7. If a faucet is leaking 30 drops per minute, with an average drop containing 0.15 milliliters of water, how much does this faucet waste in a month? [*Note.* 1000 milliliters = 1 liter = 1.057 quarts.]

8. If an office uses an average of 200 sheets of mimeograph paper each working day, and if a "ream" of 500 sheets costs $2.95, what is the monthly cost of mimeograph paper for that office?

9. Driving at 50 mph, how long would it take to travel (a) 30 miles?, (b) 100 yards?

10. Assume that you have money invested in two different money-market accounts, and you want to compare their performances. (You have neither deposited nor withdrawn any money recently.) Compute the two effective annual interest rates

(in percentages) if your balance in account (a) increases from $1,500 to $1,510.50 in 35 days, and your balance in account (b) increases from $600 to $603.60 in 28 days.

11. You are at the window watching a worker 200 yards away driving a steel stake into the ground with a sledge hammer. Assuming sound travels at 1100 feet/second, how long after you see each strike of the hammer will you hear it?

12. An employee's annual salary was increased from $11,582 to $12,103. What percentage raise did she get?

13. On September 2, when your odometer read 57,895 miles, you put 9.8 gallons of gas in your car, filling the tank to the brim. Two weeks later the odometer read 58,141 miles and it took 14.3 gallons to fill the tank. During this 2-week period (a) how many miles per gallon was your car getting? (b) how many gallons per mile was it using?

CHAPTER 2

Prime Numbers and Fractions

Given a large integer, how can one determine whether or not it is divisible by the prime numbers 2, or 3, or 5, or other numbers?

If you were asked to reduce the fraction 2279/4687 to an equivalent fraction in lowest terms, how would you proceed? (As it turns out, both the numerator and denominator are divisible by 43 but not by anything smaller.)

When is a decimal number equivalent to a fraction, and how does one find that fraction? Are there decimals which are *not* equivalent to fractions?

These are some of the questions to be answered in the present chapter.

2.1. Prime Numbers and Factorization

An integer greater than one is called a **prime number** if it has no integer factors (or divisors) other than 1 and itself. Thus the first five primes are 2, 3, 5, 7, 11.

If an integer greater than 1 is not prime, then it can be factored into a product of prime numbers (some of the prime factors possibly being repeated) called its prime factorization. Thus

$$15 = 3 \times 5 \qquad \text{and} \qquad 168 = 2^3 \times 3 \times 7$$

are the prime factorizations of 15 and 168. Confirm these by multiplication.

Example 1. Find the prime factorization of 171 (or prove that it is prime itself).

Solution. Certainly 171 is not divisible by 2. Try dividing by 3, and you will find $171 = 3 \times 57$. Now note that $57 = 3 \times 19$, and 19 is a prime number. So the prime factorization is

$$171 = 3^2 \times 19.$$

To facilitate factoring, it is useful to have simple tests for divisibility.

An integer is divisible by 2 if and only if it is even, and this is the case if and only if the last digit (on the right) is even. Thus we see at a glance that 31,698 is divisible by 2, while 31,697 is not.

An integer is divisible by 10 if and only if the last digit is zero.

An integer is divisible by 5 if and only if the last digit is either 0 or 5. Why?

There is also an easy test for divisibility by 3 or 9. An integer is divisible by 3 (or 9) if and only if the sum of its digits is divisible by 3 (or 9).

Example 2. The integer 7485 is divisible by 3 because $7 + 4 + 8 + 5 = 24$ is divisible by 3. It is not divisible by 9.

Let us pursue this example further to show *why* the method works.

Rewrite the number 7485 as follows:

$$\begin{aligned} 7485 &= 7 \times 1000 + 4 \times 100 + 8 \times 10 + 5 \times 1 \\ &= 7(999 + 1) + 4(99 + 1) + 8(9 + 1) + 5(1) \\ &= 7 \times 999 + 7 \times 1 + 4 \times 99 + 4 \times 1 + 8 \times 9 + 8 \times 1 + 5 \times 1 \\ &= (7 \times 999 + 4 \times 99 + 8 \times 9) + (7 + 4 + 8 + 5). \end{aligned}$$

The last line shows that 7485 is the sum of a number which is clearly divisible by 9 plus the number $(7 + 4 + 8 + 5)$ which is 24. Since 24 *is* divisible by 3, so is 7485. Since 24 *is not* divisible by 9, neither is 7485.

This test could have been used on the number 171 in Example 1. Since $1 + 7 + 1 = 9$, one concludes that 171 is divisible by 9.

Example 3. Find the prime factorization of 9828.

Solution. Since the last digit, 8, is even, the entire number is even. One finds

$$9828 = 2 \times 4914.$$

But 4914 is also even, and this leads us to discover

$$9828 = 2 \times 2 \times 2457.$$

Now 2457 is not divisible by 2. However, since $2 + 4 + 5 + 7 = 18$, 2457 is divisible by 9. In fact

$$9828 = 2^2 \times 9 \times 273.$$

Similarly, the fact that $2 + 7 + 3 = 12$ shows that 273 is divisible by 3. One finds

$$9828 = 2^2 \times 3^3 \times 91.$$

Since 91 is clearly not divisible by 2, 3, or 5, try dividing by 7. This yields

$$9828 = 2^2 \times 3^3 \times 7 \times 13.$$

But 13 is a prime number. So the prime factorization is complete.

Example 4. Find the prime factorization of 131.

Solution. One easily determines that 131 is not divisible by 2, 3, or 5. Next try 7 and find that this too is *not* a factor (or divisor) of 131.

There is no need to try 8, 9 or 10 since these are not primes; 8 and 10 could not be factors since 2 is not, and 9 could not divide 131 since 3 does not.

Try 11 to find that it is *not* a factor of 131.

Now there is no need to go further. The next prime is 13, and this cannot be a factor of 131 since $13^2 = 169 > 131$ (or $13 > \sqrt{131}$). So if 13 were to divide 131, the other factor would have to be less than 13. And we have already ruled out all primes below 13.

The conclusion is that 131 must itself be prime.

In general, if you are testing a given positive integer n for divisibility by primes in *increasing order*

$$2, 3, 5, 7, \ldots,$$

there is no need to test any prime p for which

$$p^2 > n.$$

If there are no prime numbers $\leq \sqrt{n}$ which divide n, then n is prime.

In order to reduce a given fraction a/b to lowest terms, one would like to divide both numerator and denominator by their "greatest common factor" (or divisor).

Example 5. Find the greatest common factor of 10,530 and 9828, and use this information to reduce 10,530/9828 to lowest terms.

Solution. From Example 3,

$$9828 = 2^2 \times 3^3 \times 7 \times 13.$$

Next factor 10,530. The process can be outlined as follows:

$$\begin{aligned} 10{,}530 &= 10 \times 1053 \\ &= 2 \times 5 \times 9 \times 117 \\ &= 2 \times 5 \times 9 \times 9 \times 13 \\ &= 2 \times 3^4 \times 5 \times 13. \end{aligned}$$

Now, in order to find the greatest common factor of two integers, one simply multiplies together the highest powers of all primes which divide into *both* of their prime factorizations.

Thus the greatest common factor of 10,530 and 9828 is

$$d = 2 \times 3^3 \times 13 \ (= 702).$$

Now reduce the original fraction to lowest terms by dividing both numerator and denominator by d:

$$\frac{10530}{9828} = \frac{2 \times 3^4 \times 5 \times 13}{2^2 \times 3^3 \times 7 \times 13} = \frac{3 \times 5}{2 \times 7} = \frac{15}{14}.$$

Before leaving Example 5, let us mention another application of prime factorization. The "least common multiple" of two integers is the product of the highest powers of all primes which occur in the prime factorization *of either* integer.

Thus, the least common multiple of 10,530 and 9828 is found, using the prime factorizations in Example 5, to be

$$2^2 \times 3^4 \times 5 \times 7 \times 13 \quad (= 147{,}420).$$

In order to add or subtract two fractions, say

$$\frac{a}{b} \pm \frac{c}{d},$$

one must first convert the fractions to equivalent ones with a common denominator. One possible common denominator is bd. But it is better to use the least common multiple of b and d if that is smaller than bd.

Example 6. Add 51/126 and 17/198.

Solution. Instead of just blindly multiplying 126 and 198, first factor each:

$$126 = 2 \times 3^2 \times 7 \qquad \text{and} \qquad 198 = 2 \times 3^2 \times 11.$$

Thus, the least common multiple of 126 and 198 is

$$2 \times 3^2 \times 7 \times 11 = 1386.$$

Letting this be the common denominator, one finds

$$\frac{51}{126} + \frac{17}{198} = \frac{51 \times 11 + 17 \times 7}{1386} = \frac{561 + 119}{1386} = \frac{680}{1386} = \frac{340}{693}.$$

(The problem could be done, but would be much more difficult, if one used for the common denominator $126 \times 198 = 24{,}948$.)

Theorem (Euclid, ca. 300 B.C.). *There are infinitely many prime numbers.*

PROOF. Suppose (for contradiction) that the primes consist of $p_1, p_2, p_3, \ldots, p_n$ and no others. Consider the integer

$$q = (p_1 p_2 \cdots p_n) + 1.$$

This integer q is not divisible by p_1, or $p_2, \ldots$, or p_n. Why?

Thus q must be another prime number. This contradiction completes the proof. □

Note that the numbers 3 and 5 are prime, and yet $3 \times 5 + 1 = 16$ is not prime. Why doesn't this destroy the above proof?

PROBLEMS

1. List the first 10 prime numbers.
2. What is the largest pair of consecutive integers which are both prime?
3. Find the prime factorization of each of the following (or show that it is already prime):
   (a) 1863 (b) 5304 (c) 6875 (d) 283 (e) 187 (f) 7485.
4. Show that an integer is divisible by 4 if and only if its last two digits form an integer divisible by 4.
5. When will an integer be divisible by 8? Justify your answer.
6. Reduce each of the following fractions to lowest terms:
   (a) 208/376 (b) 540/504
7. Perform the following addition and subtraction, doing as little work as possible.
   (a) $\dfrac{57}{208} + \dfrac{23}{376}$ (b) $\dfrac{13}{540} - \dfrac{17}{504}$.
8. If you were asked for the prime factorization of 38,009, you might conceivably have to test all primes as possible factors up to and including what?
9. Here is a test for divisibility by 11. First you must convince yourself that the integers

   $$11, \quad 99, \quad 1001, \quad 9999, \quad 100{,}001, \quad \ldots$$

   are all divisible by 11. (Note that $1001 = 990 + 11$, $100{,}001 = 99{,}990 + 11$, etc.)

   Now let us illustrate the test by considering the integer 3927. Rewrite this integer as

   $$\begin{aligned} 3927 &= 3(1001 - 1) + 9(99 + 1) + 2(11 - 1) + 7(1) \\ &= (3 \times 1001 + 9 \times 99 + 2 \times 11) + (-3 + 9 - 2 + 7). \end{aligned}$$

   The last line shows that 3927 is the sum of a number which is clearly divisible by 11 plus the number $(-3 + 9 - 2 + 7)$ which is 11. Thus 3927 is divisible by 11.

   From this example, try to determine what the correct statement should be for a test of divisibility by 11.

   Use your test to determine whether each of the following is or is not divisible by 11:
   (a) 9548 (b) 8548 (c) 57,349 (d) 9,580,736.

2.2. Greatest Common Factor

The reduction of 10,530/9828 to lowest terms was manageable in Example 5 of Section 2.1 because the numerator and denominator contained small prime factors (such as 2, 3, 5) which are quickly tested. Thus it was not difficult to factor 10,530 and 9828 and thus to find their greatest common factor, 702.

But suppose the given fraction had been 2279/4687. It would be a long and tedious process to test both the numerator and denominator for divisibility by 2, 3, 5, 7, etc. Try it if you like.

The trouble is that the smallest factor of *either* the numerator or the denominator turns out to be 43.

Fortunately, there is a systematic and efficient method for finding the greatest common factor (or divisor) of *any* two positive integers. In fact, as you will see, it is generally easier to find the common factors of two large integers than it is to factor either number separately.

Before considering this method, one should review the process of division of one integer by another.

Example 1. Divide 368 by 17, carrying out the "long division" just until the remainder becomes an integer less than 17, and interpret the result.

Solution. The division computation $368 \div 17$ is conveniently performed as follows:

$$\begin{array}{r} 21 \\ 17\overline{)368} \\ \underline{34} \\ 28 \\ \underline{17} \\ 11 \end{array}$$

Stopping at this point, the quotient and the remainder are both integers, and the remainder is less than 17 (the divisor).

The above calculation says that

$$\frac{368}{17} = 21 + \frac{11}{17}.$$

Now multiply both sides by 17 to rewrite this as

$$368 = 17 \times 21 + 11.$$

This last equation is the interpretation we will want.

Example 1 has the following generalization.

The Division Algorithm. *Let a and b be any two positive integers. Then there exist integers q (for quotient) and r (for remainder) such that*

$$a = bq + r \qquad \text{and} \qquad 0 \le r < b.$$

PROOF. Begin with the "long division" process for $a \div b$. This can be represented as

$$\begin{array}{r|l} & \underline{\ q\ } \\ b & \ a \\ & \underline{qb} \\ & \ \ r \end{array}$$

which means

$$\frac{a}{b} = q + \frac{r}{b}.$$

Here the "long division" is carried out just until the remainder r is an integer less than b. Thus q will also be an integer.

Compare all this with Example 1.

Finally, multiplying both sides of the last equation by b yields

$$a = bq + r \qquad \text{where} \qquad 0 \le r < b. \qquad \square$$

Of course, if you were trying to factor the integer a, you would be *hoping* for a zero remainder.

Now let us tackle the problem raised at the beginning of this section.

Example 2. Find the greatest common factor of the integers 2279 and 4687. (Then it will be easy to reduce the fraction 2279/4687 to lowest terms.)

Solution. Without asking why, apply the division algorithm to $4687 \div 2279$. You should find

$$\frac{4687}{2279} = 2 + \frac{129}{2279},$$

which gives

$$4687 = 2279 \times 2 + 129.$$

Now, remember, we are seeking the common factors of 2279 and 4687. But the last equation shows that any common factor of these two numbers must also be a factor of 129. (Can you see why?) Note also, from the same equation, that any common factor of 129 and 2279 must also be a factor of 4687.

So it will suffice to find the common factors of 129 and 2279—a less formidable matter than the original problem.

Next apply the division algorithm to $2279 \div 129$, and find

$$2279 = 129 \times 17 + 86.$$

By reasoning as above, it now suffices to find the common factors of 86 and 129.

Continue with this approach, computing $129 \div 86$. This yields

$$129 = 86 \times 1 + 43.$$

So now you need only consider the integers 43 and 86.

But $86 = 43 \times 2 + 0$. Thus 43 is a common factor of 43 and 86—indeed it is the greatest common factor.

Working backwards through the above arguments, 43 is the greatest common factor of 86 and 129, hence also of 129 and 2279, and hence of 2279 and 4687.

The original fraction can now be put in lowest terms.

$$\frac{2279}{4687} = \frac{2279/43}{4687/43} = \frac{53}{109}.$$

The basic technique, used repeatedly in Example 2, is described more generally as follows.

The Euclidean Algorithm for Common Factors. *Let a and b be two integers with $a > b$. Apply the division algorithm to find integers q and r such that*

$$a = bq + r \qquad \text{where} \qquad 0 \leq r < b.$$

Then any factor of a and b is a factor of b and r, and conversely.

PROOF. If d is a factor of both a and b, then d is also a factor of

$$r = a - bq.$$

Hence d is a factor of both b and r.

Conversely, if d is a factor of both b and r, then d is also a factor of

$$a = bq + r.$$

Hence d is a factor of both a and b. □

Example 3. Reduce the fraction 10,530/9828 to lowest terms.

Solution. This is the same question as in Example 5 of Section 2.1. But now let us use the new method.

The division algorithm gives

$$10{,}530 = 9828 \times 1 + 702.$$

So the common factors of 9828 and 10,530 are the same as the common factors of 702 and 9828.

Apply the division algorithm to the latter numbers to find

$$9828 = 702 \times 14 + 0.$$

Thus 702 is the greatest common factor of 702 and 9828, and hence of 9828 and 10,530.

Accordingly,

$$\frac{10{,}530}{9828} = \frac{10{,}530/702}{9828/702} = \frac{15}{14}.$$

It may seem paradoxical that factoring a single large integer can be very difficult, while finding the common factors of any two integers is quite straightforward.

Factoring the integer 2279 (without the aid of a calculator) would involve a fair amount of work. But if you introduce a second integer, the problem of finding the common factors of the two integers is easy. So why not just throw in any second integer to make the problem easier? Try it!

Unless you make a lucky guess, the only common factor of 2279 and your extra integer will be 1—and that is useless.

Thus, the reason it was relatively easy to factor 2279 in Example 2 is that the extra integer provided was special. It had an interesting factor in common with 2279—not just the number 1.

The difficulty in factoring a product of two large primes (when you do not know either factor) provides the basis for a proposed method of encrypting (or coding) messages. Banks, for example, encrypt their electronic messages for transfering funds so that unauthorized persons getting access to the communication lines cannot decipher the messages or introduce their own transfer messages. Insurance companies, stock brokers, and various government agencies also need secure communication channels.

The proposed method (due to R. L. Rivest, A. Shamir, and L. M. Adelman) was described by Martin Gardner in *Scientific American*, Aug. 1977, pp. 120–124. Without explaining the method, let it just be said that the system depends upon the product of two secret large primes—perhaps each being 50 digits long. There is no danger in revealing the product of the two primes publicly, and indeed there are reasons to do this. An unauthorized receiver of an encrypted message cannot decipher it without knowing the two prime factors. And, as you will see below, factoring the known product is virtually impossible.

Example 4. How long might it take you to factor the integer 38,009 without the aid of a calculator or computer?

Solution. Unless this number contains some small prime factors, you may have to test all primes $\leq \sqrt{38{,}009}$. That is, all the primes up to 193.

To get a rough idea how long this might take, let us assume that you could perform one division per minute, and let us assume that you try every odd integer from 3 to 193 as a possible factor. In reality, of course, you only need to test the primes from 3 to 193 but, on the other hand, many of the divisions will probably take more than 1 minute each. Thus the project might take an hour or more.

In the real world, divisions are done by computers at the rate of one million or more per second. So a properly programmed computer would factor 38,009 in a fraction of a second.

But, consider a number 100 digits long which is the product of two primes each about 50 digits long. Since the computer divides so fast, let us not bother asking it to pick out just the primes for testing. Let it simply go ahead—at great speed—and test one odd integer after another as a possible factor.

A number 50 digits long means a number at least as big as 10^{49}. So the computer must carry out about $10^{49}/2$ division calculations. How long would this take at one million divisions per second? (This is part of Problem 3.)

Problems

1. Reduce to lowest terms:
   (a) 2491/5459 (b) 7259/3599 (c) 7811/11,021.
2. Consider again the problem of factoring 38,009.
   (a) Given the additional clue that 38,009 has a (nontrivial) factor in common with 26,069, now factor 38,009.
   (b) How hard do you suppose it was to make up this problem?
3. In order to factor an integer 100 digits long, a computer might have to perform $10^{49}/2$ divisions. How long would this take at the rate of (a) one million per second? (b) one hundred million per second?

2.3. Rationals and Irrationals

Any "terminating" decimal (such as 2.8 or 0.364) is equivalent to a fraction. For example, just from the meaning of decimal notation, one finds

$$2.8 = 28/10 = 14/5,$$

$$0.364 = 364/1000 = 91/250,$$

$$1.414 = 1414/1000 = 707/500.$$

However, not every fraction is equivalent to a terminating decimal.

Example 1. To find the decimal equivalent to $\frac{1}{3}$, proceed by long division:

$$\begin{array}{r} 0.33\ldots \\ 3\overline{)1.0} \\ \underline{9} \\ 10 \\ \underline{9} \\ 1 \end{array}$$

Thus $\frac{1}{3} = 0.333\ldots$, where the three dots indicate an endless sequence of threes. (Sometimes one writes $\frac{1}{3} = 0.\overline{3}$, where the bar over the 3 means that the 3s are repeated indefinitely.)

Example 2. Convert 15/11 to an equivalent decimal.

Solution. Long division yields

$$\begin{array}{r|l} & 1.3636\ldots \\ \hline 11 & 15 \\ & 11 \\ \cline{2-2} & 4.0 \\ & 3.3 \\ \cline{2-2} & 70 \\ & 66 \\ \cline{2-2} & 4 \end{array}$$

Note that as soon as a remainder (in this case 4) repeats itself, then the computation repeats itself. Thus, without going any further in the long division, you see that $15/11 = 1.363636\ldots$ or $1.\overline{36}$. (Either notation indicates endless repetition of the pair of digits, 36.)

These two examples suggest that any fraction, a/b, is equivalent to a repeating decimal. (A "terminating" decimal is a special case of this in which the repeating pattern is just 0.) The reason for making this assertion is that when dividing integer a by integer b, the only possible remainders are 0, 1, 2, $\ldots, b - 1$. (Why?) So *eventually some remainder must repeat itself*. And, from then on, a sequence of digits in the decimal quotient repeats. See Problem 5.

The converse is also true. A repeating decimal is always equivalent to a fraction.

The next two examples should convince you of the validity of this statement.

Example 3. Convert the repeating decimal $0.\overline{135}$ to an equivalent fraction (if possible).

Solution. Let $x = 0.135135135\ldots$. Note that the repeating pattern is exactly *three digits long*. Multiply x by that power of 10 which will move the decimal point exactly *three places* to the right. In other words, multiply by 1000. Then compute $1000x - x$ as follows:

$$\begin{array}{rl} 1000x = & 135.135135135\ldots \\ x = & 0.135135135\ldots \\ \hline 999x = & 135.000000000\ldots \end{array}$$

Thus, $x = 135/999 = 15/111 = 5/37$.

Notice the reason for multiplying x by 1000—to shift the decimal point exactly the length of the repeating pattern. It is this that makes the repeating pattern cancel out when one subtracts x from $1000x$. Nothing would have been achieved by computing $10x - x$, or $100x - x$, or $10{,}000x - x$. Try it!

Example 4. Convert $3.1\overline{2}$ to an equivalent fraction.

Solution. The repeating pattern now is just *one digit long*. So, if $x = 3.12222\ldots$, one should now multiply by just 10 in order to shift the decimal point *one place*. Then $10x - x$ will be nonrepeating:

$$\begin{array}{r} 10x = 31.22222\ldots \\ x = 3.12222\ldots \\ \hline 9x = 28.1 \end{array}$$

Thus $x = 28.1/9 = 281/90$.

In Example 4 it sufficed to multiply x by 10, shifting the decimal point just one place to the right, since this made $10x - x$ a terminating decimal. However, in this case one *could* have found the answer by computing instead $100x - x$, or $1000x - x$, etc. See Problem 2. Such an approach involves a little more work than the $10x - x$ computation, and one generally prefers the shortest method.

Definition. Any number which can be expressed as a fraction is called a **rational** number. (Thus any repeating decimal is a rational.) A number which is not equivalent to a fraction is said to be **irrational**.

Example 5. The numbers $1.01001000100001\ldots$ and $9.343344333444\ldots$ are irrational since they are nonrepeating decimals. (They cannot be equivalent to fractions.)

Problems

1. Convert each of the following to an equivalent decimal:
   (a) 3/8 (b) 1/9 (c) 13/11 (d) 4/7 (e) 45/37.
2. Redo Example 4 by computing $100x - x$ instead of $10x - x$.
3. Convert each of the following to an equivalent fraction in lowest terms.
   (a) $0.\overline{81}$ (b) $1.\overline{1}$ (c) $24.1\overline{09}$ (d) $0.\overline{837}$ (e) $0.\overline{142857}$ (f) $0.\overline{9}$.
4. Describe two irrational numbers other than the ones in Example 5.
5. If a fraction a/b is converted to a decimal (by long division) what is the *maximum possible* length of the repeating pattern?

CHAPTER 3

The Pythagorean Theorem and Square Roots

By about 2000 B.C. the Egyptians knew (or believed) that if a right triangle had legs of lengths 3 and 4 units, then the hypotenuse—the side opposite the right angle—had a length of 5 units.

Later this was generalized by the Babylonians. Babylonian clay tablets dating back to about 1600 B.C. record the theorem $a^2 + b^2 = c^2$, relating the lengths of the sides of *any* right triangle. But the Babylonian mathematicians probably could not prove it.

3.1. The Theorem

This theorem is used today, as it was in ancient times, for laying out square corners of fields and foundations. It is also of vital importance in problems ranging from carpentry and navigation to astronomy and to studies of the very nature of space and time.

The Pythagorean Theorem. *If a and b are the lengths of the two legs of a right triangle and c is the length of the hypothenuse, then*

$$a^2 + b^2 = c^2.$$

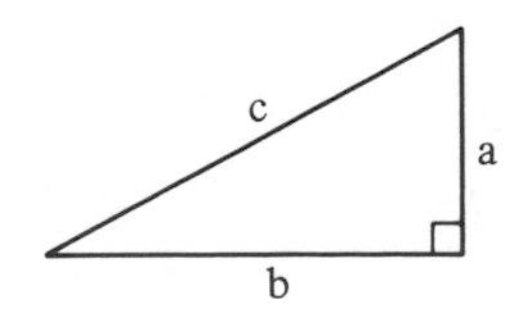

As noted above, this theorem was stated without proof around 1600 B.C. But, the first general proof is usually credited to the Greek philosopher Pythagoras (ca. 580–500 B.C.). And the theorem is named after him. (In fact, a real proof of the Pythagorean Theorem may not have been given until Euclid's, 200 years later.)

Many different proofs have been given since then. One of the simplest, described below, was discovered by U.S. Representative James A. Garfield (1831–1881) 5 years before he became the 20th President of the United States.

Garfield's proof makes use of the expression for the area of a right triangle. The triangle in Figure 1 has area $= \frac{1}{2}ab$ since it is just half of a rectangle with sides of lengths a and b.

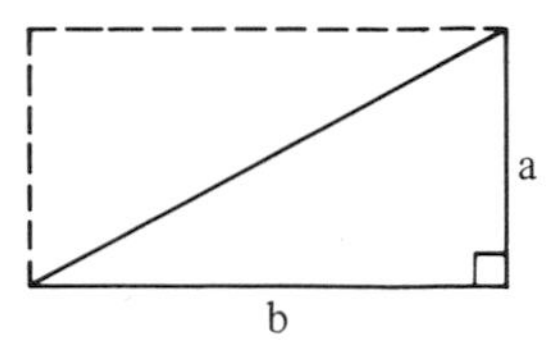

Figure 1.

Proof of the Pythagorean Theorem. Figure 2 shows three triangles forming half of a square with sides of length $a + b$. Two of these triangles (shaded) are congruent to our original right triangle. Now note that angles A, B, and D satisfy the relations

$$A + B = 90° \qquad \text{and} \qquad A + B + D = 180°.$$

So $D = 90°$, and the third triangle is also a right triangle.

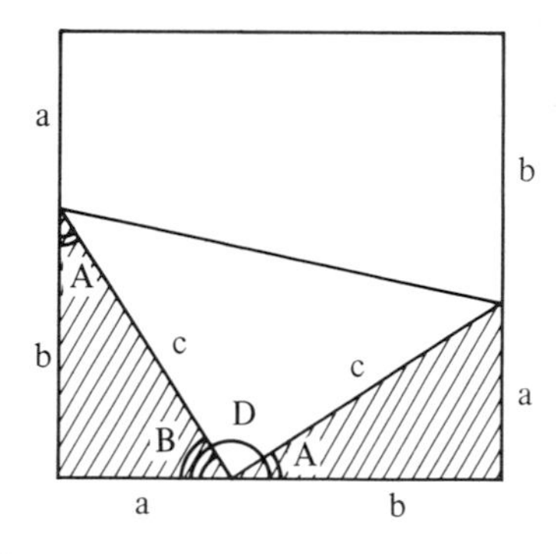

Figure 2.

The area of the half square is

$$\tfrac{1}{2}(a + b)^2 = \tfrac{1}{2}(a^2 + 2ab + b^2),$$

while the equivalent total area of the three triangles is

$$\tfrac{1}{2}ab + \tfrac{1}{2}c^2 + \tfrac{1}{2}ab.$$

Equating these two expressions, one finds

$$a^2 + 2ab + b^2 = ab + c^2 + ab \qquad \text{or} \qquad a^2 + b^2 = c^2. \qquad \square$$

Example 1. If a TV screen were rectangular with sides of lengths 9 inches and 12 inches, what would be its advertised size—the length of the diagonal?

Solution. Letting c be the length of the diagonal in inches, one has $c^2 = a^2 + b^2 = 9^2 + 12^2 = 225$. So $c = \sqrt{225} = 15$.

The following theorem provides a practical method for constructing right angles.

Converse of the Pythagorean Theorem. *If a triangle has sides of lengths a, b, and c where $a^2 + b^2 = c^2$, then it must be a right triangle. The right angle is formed at the vertex where the sides of lengths a and b meet.*

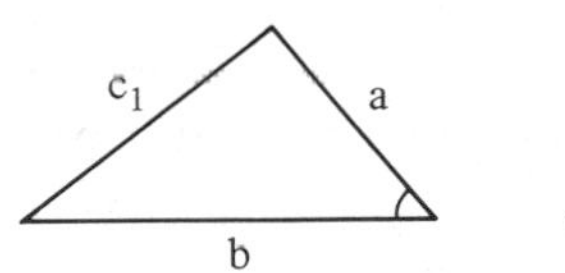

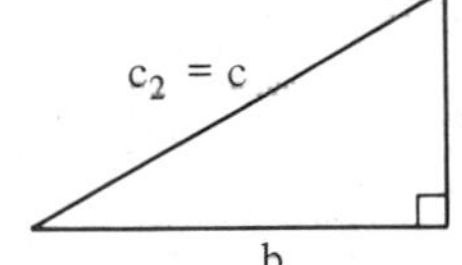

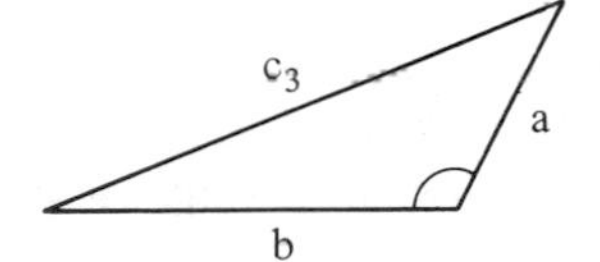

Figure 3.

PROOF. Consider the three triangles in Figure 3. Each has a side of length a and a side of length b, with the angle formed by these two sides being, respectively, an acute angle ($<90°$), a right angle, and an obtuse angle ($>90°$). Denote the lengths of the third sides in these three cases by c_1, c_2, and c_3, respectively.

In Case 2, the Pythagorean Theorem asserts that $c_2^2 = a^2 + b^2$. Thus $c_2 = c$.

In Case 1, we must have $c_1 < c_2$ since the angle is acute and the side lengths a and b are unchanged. Hence $c_1^2 < a^2 + b^2$, and this kind of triangle is ruled out.

Similarly, in Case 3, $c_3 > c_2$, so $c_3^2 > a^2 + b^2$.

Only the right triangle satisfies $c^2 = a^2 + b^2$. $\square$

Example 2. Describe a method for laying out the 90° corner of a foundation for a house.

Solution. Create a triangle by measuring out sides of lengths 3, 4, and 5 yards. Then, since $3^2 + 4^2 = 5^2$, one must have a right angle where the sides of lengths 3 and 4 meet.

This idea was used by the Egyptians as far back as 2000 B.C. for laying out square corners. They did not even need a standard unit of length (such as a yard or meter). All that was needed was enough rope with knots, or other markers, at uniformly spaced intervals (see Figure 4). Then a right angle would be created by forming a triangle with sides of lengths 3, 4, and 5 spaces between knots (or 6, 8, and 10 spaces, etc.).

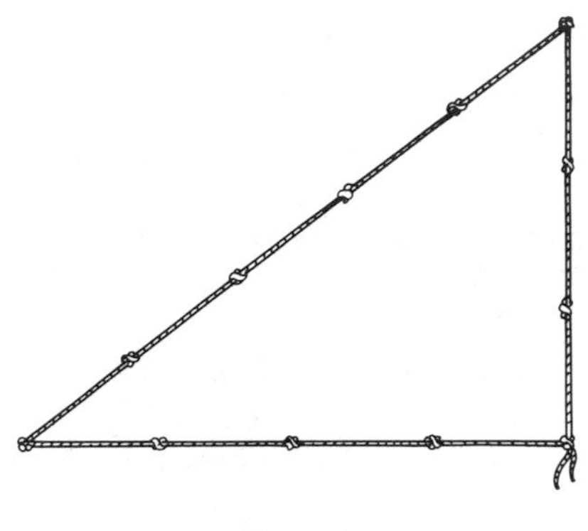

Figure 4.

Note. When you need the square root of a large number or of a fraction or a decimal, it sometimes helps to use the facts that

$$\sqrt{ab} = \sqrt{a}\sqrt{b} \qquad \text{and} \qquad \sqrt{\frac{a}{b}} = \frac{\sqrt{a}}{\sqrt{b}}.$$

These are proved by calculating

$$(\sqrt{a}\sqrt{b})^2 = (\sqrt{a})^2(\sqrt{b})^2 = ab \qquad \text{and} \qquad \left(\frac{\sqrt{a}}{\sqrt{b}}\right)^2 = \frac{(\sqrt{a})^2}{(\sqrt{b})^2} = \frac{a}{b}.$$

Example 3

$$\sqrt{14400} = \sqrt{144}\sqrt{100} = 12 \times 10 = 120,$$

$$\sqrt{\frac{25}{16}} = \frac{\sqrt{25}}{\sqrt{16}} = \frac{5}{4},$$

$$\sqrt{1.44} = \sqrt{\frac{144}{100}} = \frac{\sqrt{144}}{\sqrt{100}} = \frac{12}{10} = 1.2.$$

Although it is easy to find $\sqrt{14400}$, the same *cannot* be said of $\sqrt{1440}$. The trouble is that 1440 is not a product of two integer squares.

PROBLEMS

1. If the legs of a right triangle are $a = 12$ and $b = 5$, find the length of the hypotenuse.
2. If one leg of a right triangle has length 8 and the hypotenuse has length 17, find the length of the other leg.
3. Find the length of a guy wire from the top of a tent pole 15 feet tall to a point on the ground 20 feet from the base of the pole.
4. If you were laying out a rectangle 24 by 32 feet for the foundation of a house, what should be the length of the diagonals—to assure that your quadrilateral is really a rectangle?
5. A small boy is flying his kite at the end of 260 feet of string. Another small boy standing directly beneath the kite is 100 feet from the first boy. How high is the kite?
6. Given two 2-foot lengths of string and a third piece 3 feet long, could you form a triangle using these as the sides? If so, what kind of an angle would be formed where the 2-foot sides meet? (Acute, right, or obtuse.)
7. Given three pieces of string of lengths 5 meters, 8 meters, and 9 meters, could you form a triangle using these as the sides? If so, what kind of angle would be found where the 5-meter and 8-meter sides meet?
8. A picture is mounted on a thin but stiff rectangular board 3 feet by 5 feet. Could it pass through a rectangular opening 1 foot 9 inches by 2 feet 6 inches?
9. Find quickly (in your head if possible) (a) $\sqrt{12{,}100}$, (b) $\sqrt{1.21}$, (c) $\sqrt{640{,}000}$, (d) $\sqrt{0.0064}$, (e) $\sqrt{64/81}$
10. The Great Pyramid of the Pharoah Khufu (Cheops) is more than 750 feet on a side, and yet its corners are right angles to within an error of 1/300 of a degree. But this pyramid was built during the 2600's B.C.—presumably hundreds of years before the Pythagorean Theorem was known. How could the builders have made square corners?

3.2. Square Roots Which Are Irrational

The examples and problems in Section 3.1 were designed to work out nicely with answers that were integers.

But now consider what is perhaps the most obvious right triangle of all—one with equal legs. Specifically, let $a = b = 1$. Then, by the Pythagorean Theorem, the hypotenuse has length c where

$$c^2 = a^2 + b^2 = 1^2 + 1^2 = 2.$$

Thus, $c = \sqrt{2}$.

Now you may know that $\sqrt{2} \cong 1.4 = 14/10 = 7/5$ or $\sqrt{2} \cong 1.414 =$

$1414/1000 = 707/500$. But these rational numbers are only approximations. Indeed $(1.414)^2 = 1.999396$, so 1.414 is too small to be the exact value of $\sqrt{2}$.

In fact there is *no* rational number equal to $\sqrt{2}$.

Theorem. *$\sqrt{2}$ is irrational.*

This is not hard to prove, and the proof will be given. However, the method is quite different from that used in Section 2.3 (Example 5) to exhibit some irrationals. One cannot prove that $\sqrt{2}$ is irrational by showing that its decimal form has no repeating pattern because one does not have its complete decimal representation. There are methods for finding the decimal representation of $\sqrt{2}$ to any desired accuracy. But how would you know that a repeating pattern does not begin at some point beyond where you left off?

The proof will use a "lemma" or preliminary theorem. Both the lemma and the main result will be obtained via "proof by contradiction."

Lemma. *Let m^2 be even for some integer m. Then m itself is even.*

PROOF. Suppose (for contradiction) that m is odd. Then

$$m = 2p + 1$$

for some integer p. (That is, m is an even number plus 1.) So

$$m^2 = (2p + 1)^2 = 4p^2 + 4p + 1,$$

which is odd. But m^2 is known to be even.

This contradiction forces one to conclude that the supposition "m is odd" was false. So m must be even. □

PROOF THAT $\sqrt{2}$ IS IRRATIONAL. Suppose (for contradiction) that $\sqrt{2}$ is rational. Then

$$\sqrt{2} = m/n,$$

where m and n are positive integers. Without loss of generality assume that m and n are *not both even*. If they were, you could divide both numerator and denominator by 2. And this could be repeated until at least one of the numbers was odd. For example,

$$\frac{1414}{1000} = \frac{707}{500},$$

and you could use the second form.

Now, from the hypothesis $\sqrt{2} = m/n$, one finds $m = \sqrt{2}n$ and hence

$$m^2 = (\sqrt{2}n)^2 = 2n^2.$$

This shows that m^2 is an even number. Thus, m itself is even by the lemma. This means

$$m = 2p$$

for some integer p. But then

$$4p^2 = m^2 = 2n^2$$

or

$$n^2 = 2p^2.$$

This shows that n^2 is even, and so it follows that n is even. But that contradicts the fact that *not both* m and n are even.

Now each step above is a logical deduction *except* for the supposition that "$\sqrt{2}$ is rational," i.e., a ratio of two integers. So one is forced to conclude that $\sqrt{2}$ *cannot* be expressed as the ratio of two integers. □

Pythagoras was the founder of a society (or school) for scientific and religious thinking which became known as the brotherhood of Pythagoras. Its members are referred to as Pythagoreans. A central belief and teaching of the Pythagoreans was that everything could be related to integers or fractions. Thus it was a disaster of immense proportions when one member of the brotherhood discovered that $\sqrt{2}$, the hypotenuse of a simple right triangle, could not be described by a fraction.

Here is the difficulty. Using only a piece of rope with equally spaced knots, you could lay out a 3–4–5 right triangle. But you could never do the same for a 1–1–$\sqrt{2}$ right triangle. Since $\sqrt{2} \cong 1.4 = 7/5$, you could *approximate* such a triangle by using side lengths of 5, 5, and 7 spaces between knots. Or, since $\sqrt{2} \cong 1.41 = 141/100$ you could make a better approximation by using lengths of 100, 100, and 141. (This means either making the triangle bigger or reducing the spacing between knots.) Better yet would be lengths of 500, 500, and 707. But, no matter how small you make the uniform spacing between knots, the exact ratio of $1 : \sqrt{2}$ can never be achieved. Such numbers as 1 and $\sqrt{2}$ are said to be *incommensurable*.

According to one legend, this fact was such a threat to the teachings of the Pythagoreans that the poor fellow who made the discovery was thrown overboard at sea to drown. And the other members of the brotherhood were sworn to secrecy on his discovery.

Problems

1. If one leg of a right triangle has length 1 and the hypotenuse has length 2, how long is the other leg?

2. Prove that if m^2 is divisible by 3 for some integer m, then m itself is divisible by 3. [*Hint*. This is analogous to the lemma. Note that if m is an integer *not* divisible by 3, then *either* $m = 3p + 1$ *or* $m = 3p + 2$ for some integer p.]

3. Prove that $\sqrt{3}$ is irrational. [*Hint*. Write a proof analogous to that for $\sqrt{2}$ in the text. You will need the result of Problem 2.]

*4. Prove that $\sqrt{5}$ is irrational.

*5. What happens if you try to prove that $\sqrt{4}$ is irrational by a method analogous to that used for $\sqrt{2}$ and $\sqrt{3}$?

6. Why has no one ever written down (in feet and inches) the exact length of the diagonal of a square 10 feet by 10 feet? [*Hint*. Use the fact that $\sqrt{ab} = \sqrt{a}\sqrt{b}$.]

3.3 Computation of Square Roots by Successive Approximation

It can be shown that the square root of any positive integer is either an integer or an irrational number. Thus, the square root of each positive integer except 1, 4, 9, 16, 25, 36, ... will be irrational. Most other positive rational numbers (e.g. 1/2, 17/4, and 2.3) also have *irrational* square roots.

Even though one cannot express an irrational number exactly as a terminating or repeating decimal (or a fraction), there is a straightforward procedure for approximating the square root to any accuracy desired. This is the method of "successive approximations."

Suppose you are interested in finding $\sqrt{a}$ for some $a > 0$. Make a *guess* of some number $x > 0$ as an approximation to $\sqrt{a}$. If x were exactly $\sqrt{a}$, then $a = x^2$ and so $a/x = x$.

But, most likely the guessed value of x will not equal $\sqrt{a}$. It will be either too large or too small. Note that

$$\text{if } x > \sqrt{a}, \text{ then } a < x\sqrt{a} \text{ so that } a/x < \sqrt{a}$$

and

$$\text{if } x < \sqrt{a}, \text{ then } a > x\sqrt{a} \text{ so that } a/x > \sqrt{a}.$$

Thus, unless $x = \sqrt{a}$ exactly, one has either

$$x < \sqrt{a} < a/x \qquad \text{or} \qquad a/x < \sqrt{a} < x.$$

So the (unknown) exact value or $\sqrt{a}$ must always lie between x and a/x, where $x > 0$ is *any* guessed value.

Example 1. Let $a = 2$, and try to learn something about the value of $\sqrt{2}$. Suppose we take a guess of $x = 1.5$. Then, since $2/1.5 = 1.\overline{3}$, it follows that

$$1.3 < \sqrt{2} < 1.5.$$

Now, with hindsight, one might think of starting over again with the guess $x = 1.4$ (the average of 1.3 and 1.5). Taking $x = 1.4$, one finds

$$1.4 < \sqrt{2} < \frac{2}{1.4} = 1.43$$

which is more useful information than was obtained from taking $x = 1.5$.

The Method of Successive Approximations *is based on the above ideas. To compute* $\sqrt{a}$, *where* $a > 0$, *begin with any first guess* $x_1 > 0$. *Then define*

$$x_2 = \left(\frac{a}{x_1} + x_1\right)\frac{1}{2}.$$

Now regard x_2 *as your guess for* $\sqrt{a}$ *and repeat the process. This means define*

$$x_3 = \left(\frac{a}{x_2} + x_2\right)\frac{1}{2},$$

and so on.

Since one of the numbers x_1 or a/x_1 is too big and the other is too small, it seems reasonable that their average x_2 will be a better approximation to $\sqrt{a}$. Hopefully, the sequence of numbers $x_1, x_2, x_3, \ldots$ provides successively better approximations to $\sqrt{a}$.

At any stage you can discover how close you are to the exact value of $\sqrt{a}$ by reasoning as in Example 1 above.

Example 2. Use this method to compute $\sqrt{2}$, starting with the initial guess $x_1 = 1$. (This is not a particularly good first guess, but that will not matter.)

Solution. Since $x_1 = 1$,

$$x_2 = \left(\frac{2}{x_1} + x_1\right)\frac{1}{2} = \left(\frac{2}{1} + 1\right)\frac{1}{2} = \frac{3}{2} = 1.5.$$

Then

$$x_3 = \left(\frac{2}{x_2} + x_2\right)\frac{1}{2} = \left(\frac{2}{1.5} + 1.5\right)\frac{1}{2} = (1.3333 + 1.5)\frac{1}{2} = 1.417$$

and

$$x_4 = \left(\frac{2}{x_3} + x_3\right)\frac{1}{2} = \left(\frac{2}{1.417} + 1.417\right)\frac{1}{2} = (1.4114 + 1.417)\frac{1}{2} = 1.4142.$$

Note that each stage of this calculation provides a check on the accuracy of the previous stage. For example, as soon as you have computed $2/1.417$ (for finding x_4) you have shown that the exact value of $\sqrt{2}$ lies somewhere between 1.4114 and 1.417. And, if you proceed to compute $2/x_4 = 2/1.4142 = 1.414227$, you will have shown that $\sqrt{2} = 1.4142$ correct to five figures.

Example 3. Find $\sqrt{3}$ correct to four figures.

Solution. This time, let $x_1 = 2$. Then

$$x_2 = \left(\frac{3}{x_1} + x_1\right)\frac{1}{2} = \left(\frac{3}{2} + 2\right)\frac{1}{2} = \frac{7}{4} = 1.75$$

and

$$x_3 = \left(\frac{3}{x_2} + x_2\right)\frac{1}{2} = \left(\frac{3}{1.75} + 1.75\right)\frac{1}{2} = (1.714 + 1.75)\frac{1}{2} = 1.732.$$

Now compute $3/x_3 = 1.7321$ to conclude that the exact value of $\sqrt{3}$ lies somewhere between 1.732 and 1.7321. So $\sqrt{3} = 1.732$, correct to four figures.

In Examples 2 and 3 just a few easy steps yielded good approximations to $\sqrt{2}$ and $\sqrt{3}$. It can be proved—although we shall not do it—that the method described here always works for finding square roots; and it works no matter what one chooses for $x_1 > 0$.

The method of successive approximations is straightforward, but it can be a little tedious. So one ought to always be alert for shortcuts. An identity which is often useful is

$$\sqrt{\frac{1}{a}} = \frac{1}{\sqrt{a}} = \frac{\sqrt{a}}{a} \qquad \text{for any } a > 0.$$

The second equality follows from

$$\frac{1}{\sqrt{a}} = \frac{1}{\sqrt{a}}\frac{\sqrt{a}}{\sqrt{a}} = \frac{\sqrt{a}}{a}.$$

Also recall the identities in the note on page 36.

Example 4. Use the values of $\sqrt{2}$ and $\sqrt{3}$ found above to obtain $\sqrt{18}$, $\sqrt{12}$, and $\sqrt{0.5}$ with a minimum of effort.

Solution

$$\sqrt{18} = \sqrt{9 \times 2} = \sqrt{9}\sqrt{2} = 3\sqrt{2} \cong 4.24,$$

$$\sqrt{12} = \sqrt{4 \times 3} = \sqrt{4}\sqrt{3} = 2\sqrt{3} \cong 3.46,$$

$$\sqrt{0.5} = \sqrt{\frac{1}{2}} = \frac{1}{\sqrt{2}} = \frac{\sqrt{2}}{2} \cong 0.707$$

Problems

1. Compute $\sqrt{2}$ starting with $x_1 = 2$.
2. Compute $\sqrt{2}$ starting with $x_1 =$ any positive number of your choice other than 1 or 2.
3. Compute $\sqrt{5}$ correct to four figures.
4. What length lumber should you buy for the rafters of a 24 by 24-foot garage if the roof slope gives a rise of 5 inches for each 12 inches horizontal? [*Note.* Lumber is usually sold only in lengths which are multiples of 2 feet.]

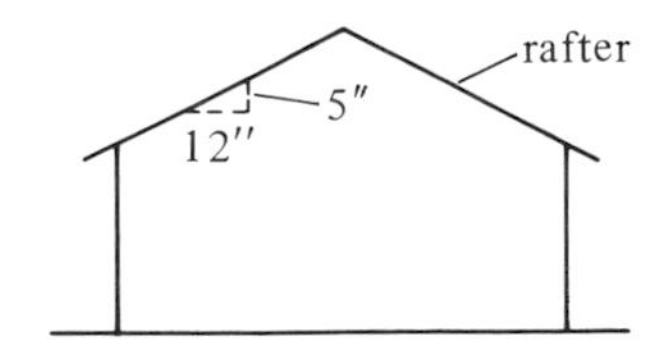

5. Use the results of Examples 2 and 3 and Problem 3 to find *quickly* (with very little further computation)
   (a) $\sqrt{200}$, (b) $\sqrt{0.03}$, (c) $\sqrt{50{,}000}$, (d) $\sqrt{1/3}$.

6. If you were laying out a rectangle for the foundation of a house 24 feet by 40 feet, what should be the length of the diagonals?

7. A ship is tethered to a dock by a taut rope 25 feet long. If the rope is attached to the bow of the ship 10 feet above the mooring ring on the edge of the dock, how far is the ship from the dock (to the nearest foot)?

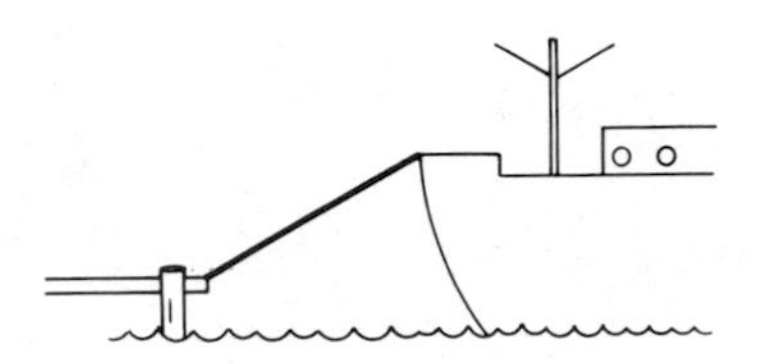

8. How long must a guy wire on a sailboat be in order to reach from the top of a 20-foot mast to a point on the bow 6 feet from the base of the mast? Give your answer rounded *upward* to the nearest foot.

*9. A fairly steep hill might rise 1 foot for each 10 feet horizontally. What percentage error would a surveyor make if he measured the distance up the hill and considered that to be the horizontal distance?

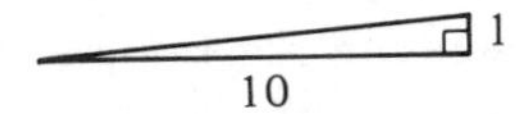

CHAPTER 4

Elementary Equations

Many day-to-day problems can be solved most easily with the aid of a little elementary algebra. Examples in this chapter will treat currency exchanges, speed and distance problems, mixtures, levers, and conversions between Celsius and Fahrenheit.

In most cases, one must first translate some given information into algebraic language, then solve the resulting algebraic equation(s). The first step—translating the "word problem" into algebraic form—seems to give many students "fits" of anxiety. The recommended procedures and the examples given in this chapter are supposed to convince you that "word problems" are not so formidable, and that *you* can handle them without anguish.

4.1. Equations in One Unknown

Each example and problem in this section can be solved via the following outline.

(1) Write a *careful and unambiguous definition of* x, the unknown. Usually x will be chosen to be the quantity, or one of the quantities, sought in the problem. Be sure to *state the units of* x, if any, in the definition.

(2) Translate the relevant information in the problem into an equation. This is done by finding some appropriate quantity which can be expressed in *two different ways*, and then equating the two expressions.

(3) Solve for x, and formulate your conclusions.

(4) Check your answer in the equation of step (2) or, better yet, check it against the wording of the original problem.

Example 1. Johnson and Son advertise a total of "25 years experience" as TV repairmen. If the elder Johnson has 4 years more experience than his son, how long has each been fixing TV sets?

Solution. Following the above outline,

(1) Let x = number of years experience of the son.

(2) Now express the *total* years of experience of father and son in two different ways. Note that $x + 4$ = number of years experience of the father. So

$$x + (x + 4) = \text{total number of years experience of both.}$$

But this was given as 25. So

$$x + (x + 4) = 25.$$

(3) This equation is easily solved for x as follows:

$$2x + 4 = 25$$

$$\therefore 2x = 21$$

$$\therefore x = 10.5.$$

So the son has 10.5 years experience and his father has $x + 4 = 14.5$ years.

(4) This solution checks against the original wording of the problem since the total experience is $10.5 + 14.5 = 25$ years, and the father has $14.5 - 10.5 = 4$ years more experience than his son.

Example 2. If a U.S. dollar is equivalent to 1.15 Canadian dollars, what is a Canadian dollar worth in U.S. money? (It is *not* 85 cents.)

Solution. Once again the steps of the outline are identified.

(1) Let x = value in U.S. dollars of one Canadian dollar.

(2) Then 1.15 Canadian dollars is worth $1.15x$ U.S. dollars. But it was stated that 1.15 Canadian dollars is worth one U.S. dollar. Thus

$$1.15x = 1.$$

(3) Solve for $x = 1/1.15 = 0.87$ (rounded off). So a Canadian dollar translates into 87 cents in U.S. money.

(4) As a check, note that if a Canadian dollar is worth 0.87 U.S. dollars, then 1.15 Canadian dollars is worth $(1.15)(0.87) = 1.0005$ (or 1.00 rounded off) U.S. dollars. The answer is confirmed.

(This example is artificial. The actual rate of exchange varies from day to day, and is published in the financial section of many newspapers.)

Example 3. A motorist drives from city A to city B, 100 miles away, at 30 mph and then returns at 50 mph. What is her average speed for the round trip? (No, it is *not* 40 mph.)

Solution

(1) Let x = average speed for the round trip in mph.

(2) Next seek two different ways of expressing the *total time* for the round trip. Since

$$\text{distance} = \text{speed} \times \text{time},$$

so

$$\text{time} = \frac{\text{distance}}{\text{speed}}.$$

Thus the trip from A to B required 100/30 hours. Similarly, the return trip took 100/50 hours. So the total trip took 100/30 + 100/50 hours. On the other hand, a round trip of 200 miles at an average speed of x mph takes $200/x$ hours. Equating the two expressions for the total time of the round trip one has

$$\frac{200}{x} = \frac{100}{30} + \frac{100}{50}.$$

(3) Solve this equation for x as follows:

$$\frac{2}{x} = \frac{1}{30} + \frac{1}{50} = \frac{5+3}{150} = \frac{8}{150}.$$

$$\therefore x = \frac{2 \times 150}{8} = 37.5.$$

Conclusion: Her average speed for the round trip is 37.5 mph.

(4) Check this by recomputing the total time for the round trip. On the one hand it is $100/30 + 100/50 = 10/3 + 2 = 16/3$ hours. While, using the average speed of 37.5 mph, the total time is $200/37.5 = 400/75 = 16/3$ hours; the answer checks.

Example 4. An automobile cooling system contains 10 liters of a mixture of water and antifreeze which is 25% antifreeze. How much of this mixture must be drained out and replaced with pure antifreeze so that the resulting 10 liters will be 40% antifreeze?

Solution

(1) Let x = number of liters to be drained out and replaced with pure antifreeze.

(2) The total amount of antifreeze finally in the system can now be computed two ways. After draining x liters of the original mixture, the amount of *antifreeze* remaining in the system is $(10 - x)25/100$. So, when x liters of pure antifreeze are added to this, the total content of antifreeze becomes

$$(10 - x)\frac{25}{100} + x \text{ liters.}$$

The amount of antifreeze needed in an automobile cooling system depends on the climate.

On the other hand, this is supposed to equal 40% of 10, or 4 liters. Thus

$$(10 - x)\tfrac{1}{4} + x = 4.$$

(3) Therefore $10 - x + 4x = 16$, which gives $3x = 6$, or $x = 2$. So one must drain and replace 2 liters.

(4) To check this, note that draining 2 liters of the original mixture leaves 8 liters of mixture containing $8 \times 0.25 = 2$ liters of antifreeze. Then, adding 2 liters of antifreeze gives a total of 10 liters containing $2 + 2 = 4$ liters of antifreeze. This is the required 40%.

The next example involves a lever. Levers enable different forces or weights to be balanced as follows. A force F_1 on a lever at a distance d_1 to the left of the fulcrum (as in Figure 1) will balance a parallel force F_2 on the lever at a distance d_2 to the right of the fulcrum if

$$F_1 \times d_1 = F_2 \times d_2.$$

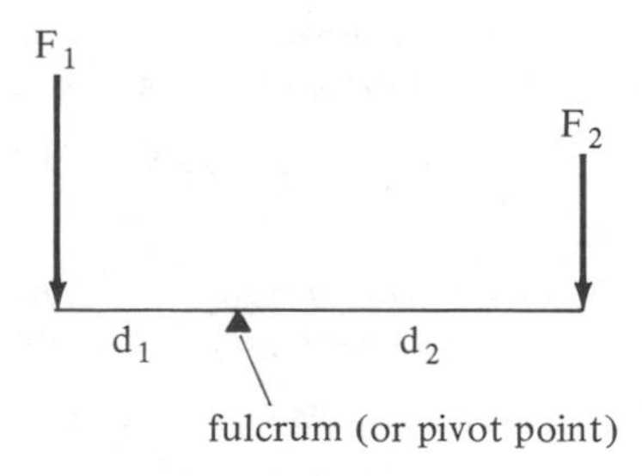

Figure 1.

Example 5. If a 120-pound person wants to balance (or raise) a 600-pound rock using an 8-foot lever, where must she put the fulcrum (perhaps another rock)? See Figure 2.

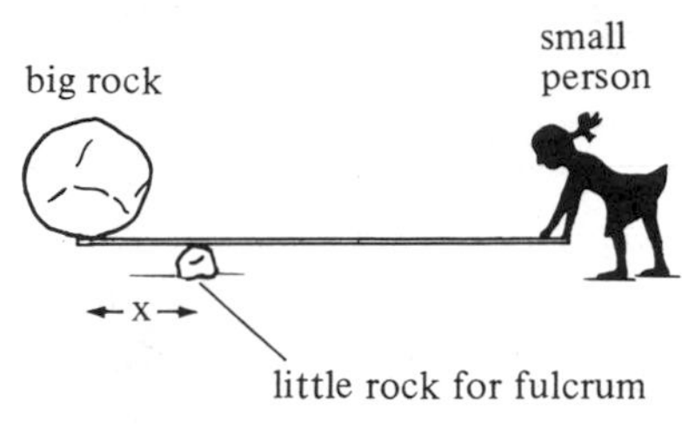

Figure 2.

Solution (with step numbers omitted).

Let $x =$ distance in feet from rock to fulcrum.

Then $8 - x =$ distance in feet from person to fulcrum. For balance, one requires

$$600x = 120(8 - x)$$

$$\therefore\ 5x = 8 - x$$

$$\therefore\ 6x = 8 \qquad \text{or} \qquad x = 4/3.$$

The fulcrum must be no more (and preferably less) than 1 foot 4 inches from the rock.

The check is left to the reader.

Problems

1. Kilroy is 5 years older than Rupert, and the sum of their ages is 37 years. How old is each?
2. Assuming a British pound is worth \$1.35, what is \$1.00 worth in pounds?
3. Suppose you have \$10,000 available for the down payment on a house, and suppose the bank requires that the down payment be at least 20% of the selling price. What limitation does this impose on the price of a house you can buy?
4. If the total cost of an item *including* a 6% sales tax is \$21.15, what was the list price of the item?
5. A 16-ounce package of cookies is said to contain "23% more than the regular size" package. How many ounces does the "regular" size contain?
6. An automobile cooling system contains eight quarts of a mixture of water and antifreeze which is 20% antifreeze. How much of this mixture must be drained out and replaced with pure antifreeze so that the resulting eight quarts will be 50% antifreeze?

7. A large barrel contains 10 liters of a mixture of water and antifreeze which is 25% antifreeze. How much antifreeze must be added so that the resulting mixture will be 40% antifreeze? (Note how this question differs from Example 4.)

8. A bottle and a cork together cost $1.10. If the bottle costs $1.00 more than the cork, what does the cork cost?

9. Your chain saw engine requires a fuel mixture that is 19 parts gasoline to 1 part oil. If you already have on hand 1 pint of a mixture of 4 parts gasoline and 1 part oil, how much gasoline should you add to get the right mixture for your chain saw?

10. A plane makes daily round-trip flights between city A and city B 400 miles west of A, always flying at a constant air speed of 200 mph. So on a calm day the round trip takes 4 hours flying time. How long (flying time) would the round trip require if there were a constant westerly wind at 40 mph?

11. A motorist sets out to drive from city A to city B, 100 km (kilometers) away. When he is halfway there he discovers that he has only been averaging 50 km/hour. If he wanted his overall average to be 60 km/hour, how fast would he have to drive for the second half of the trip?

12. A person fires a bullet at a tin can, and 2 seconds later hears the sound of the bullet hitting the can. Assuming the bullet travels at 3000 feet/second and that sound travels at 1100 feet/second, how far away is the target?

13. A motorist sets out on a trip at 40 mph. Half an hour after she leaves, her husband discovers that she forgot something, and (knowing her route) he sets forth in pursuit at 55 mph. How long will it take him to catch up with his wife?

14. What are the dimensions of a TV screen advertised as "20-inch" (diagonal)? [*Note.* A TV screen is approximately a rectangle with height equal to three-fourths of its width.]

15. A rope hanging from the top of a flag pole reaches the ground with 1 foot to spare. When pulled taut, the rope just reaches the ground at a point 11 feet from the base of the pole. How tall is the flag pole?

16. If a rope hanging from the top of a 100-foot tall pole reaches the ground with $1\frac{1}{2}$ inches to spare, at what distance from the base of the pole will it just touch the ground when pulled taut? [*Hint.* The actual equation you should find here can be solved most efficiently with the aid of Example 7 of Section 1.1. (See also Problem 7 of that section).]

17. A seesaw consists of a board 18 feet long balanced on a pivot (or fulcrum) at its midpoint. If a big boy weighing 120 pounds wants to balance his little brother weighing 50 pounds who is sitting on one end of the seesaw, where should the big boy sit?

18. A balance scale is built as follows. A 25-inch bar rests on a pivot point just 1 inch from one end of the bar. The load to be weighed hangs from the end of the bar near the pivot point, and the weights are hung from the other end. If a 200-pound object is put on the scale, what weight is required on the other end to balance it? See Figure 3.

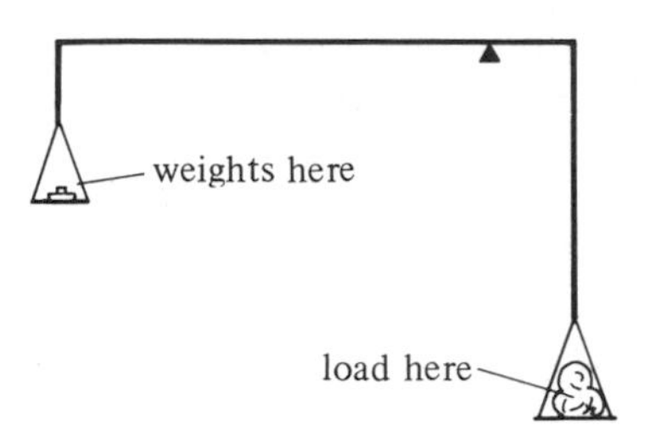

Figure 3.

19. Regarding Example 5, why can you not move an arbitrarily heavy rock merely by making x sufficiently small?

20. A man is five times as old as his son. In 7 years he will be only three times as old as his son. How old are they now?

21. Hortense is three times as old as her daughter, but in 10 years she will be only twice as old as her daughter. How old are they now?

4.2. The Use of Two or More Unknowns

Sometimes it is convenient to assign names to more than one unknown quantity, for example, to use x and y (or x, y, and z) to represent unknowns, instead of just x. Then, in order to solve, you generally need as many equations as you have unknowns.

Often you can choose to solve a particular "word problem" using either one unknown or two.

Four steps should be followed, as in the previous section.

(1) Write *careful and unambiquous definitions of the unknowns, stating the units if any.*

(2) Translate the relevant information in the statement of the problem into equations involving the unknowns you defined. (Again this is done by finding appropriate quantities which can be expressed in two different ways, and then equating the two expressions.)

(3) Solve the system of two or more "simultaneous equations" for the unknowns, say x and y. You do this by combining the equations of step (2) in such a way as to get just one equation in one unknown. (You may multiply both sides of an equation by the same number and you may add or subtract the respective sides of two or more equations. The examples below will refresh your memory about this.)

(4) Finally, check your answers against the original problem.

Example 1. Consider again Johnson and Son's claim of "25 years experience" as TV repairmen, with the elder Johnson having 4 more years experience than his son. How long has each been in the business?

Solution (using two unknowns)

(1) Let x = number of years experience of the son.
Let y = number of years experience of the father.

(2) Now, since their total experience is 25 years,

$$x + y = 25.$$

And, since the father has 4 more years experience than his son,

$$y = x + 4.$$

(3) Here is one method of solving this pair of equations for the unknowns x and y. Rewrite the equations as

$$y + x = 25,$$

$$y - x = 4.$$

Now add the left-hand sides and the right-hand sides to obtain

$$2y = 29$$

$$\therefore y = 14.5.$$

Use this information in, say, the second original equation to find $x = y - 4 = 14.5 - 4 = 10.5$. So Mr. Johnson has 14.5 years experience and his son has 10.5 years.

(4) The check is exactly as it was in Example 1 of Section 4.1.

Example 2. A certain investor has stock in Company A and Company B worth a total of \$10,000. He mentions that these investments pay him \$855 per year in dividends. You look in the newspaper and find that the stock of Company A is worth \$24.00 per share and pays an annual dividend of \$2.50 per share, while Company B is worth \$16.00 per share and pays \$1.20 per share annually. How many shares of each company does the investor own?

Solution

(1) Let x = number of shares he owns of Company A.
Let y = number of shares he owns of Company B.

(2) Then the total value of his shares of Company A is \$$24x$ and the total value of his holdings of Company B is \$$16y$. So his total in the two companies is \$$24x + 16y$. Hence

$$24x + 16y = 10{,}000.$$

His total annual dividend from Company A is \$$2.5x$ and from Company B is \$$1.2y$. But these must add up to \$855. So a second equation is

$$2.5x + 1.2y = 855.$$

(3) To solve, simultaneously, the pair of equations in step (2), multiply both sides of the first equation by 3 and multiply both sides of the second equation

by 40. The equations then become

$$72x + 48y = 30{,}000$$

and

$$100x + 48y = 34{,}200.$$

This choice of multipliers has resulted in a pair of equations with matching coefficients in front of y. Now, in this latest pair of equations, subtract the first from the second to find

$$28x = 4200 \qquad \text{(one equation in one unknown).}$$

Thus $x = 4200/28 = 150$. Now put this value of x into either one of the original equations in step (2), say the first. This gives

$$24(150) + 16y = 10{,}000$$

$$\therefore\ 16y = 10{,}000 - 3600 = 6400$$

$$\therefore\ y = 6400/16 = 400.$$

So the investor holds 150 shares of Company A and 400 shares of Company B.

(4) To check, note that the total value of his holdings in the two companies would then be $150(24) + 400(16) = 3600 + 6400 = 10{,}000$ dollars. And his total annual dividend income from the two would be $150(2.50) + 400(1.20) = 375 + 480 = 855$ dollars, as required.

It would have been harder (but not impossible) to treat Example 2 using just one unknown.

Example 3. You have two bags of fertilizer. One is 12% nitrogen and the other is 20% nitrogen. How much of each should you mix to produce 10 pounds of fertilizer containing 15% nitrogen?

Solution (without numbering the steps)

Let x = number of pounds needed of 12%-nitrogen fertilizer
and let y = number of pounds needed of 20%-nitrogen fertilizer.

Then the total amount of the mix will be

$$x + y = 10,$$

and the total content of nitrogen will be

$$0.12x + 0.2y = 0.15(10).$$

Rewrite the first of these equations unchanged, and the second multiplied by 5. Thus

$$x + y = 10$$

and

$$0.6x + y = 7.5.$$

Now subtract to find

$$0.4x = 2.5$$

$$\therefore x = 2.5/0.4 = 25/4.$$

So one needs $6\frac{1}{4}$ pounds of the 12%-nitrogen fertilizer and, from the equation $y = 10 - x$, $3\frac{3}{4}$ pounds of the 20%-nitrogen fertilizer.

To check this, note that $6\frac{1}{4} + 3\frac{3}{4} = 10$ pounds, as it should. And the total nitrogen content of the mixture will be

$$\left(6\frac{1}{4}\right)(0.12) + \left(3\frac{3}{4}\right)(0.2) = \frac{25}{4}(0.12) + \frac{15}{4}(0.2)$$

$$= \frac{3}{4} + \frac{3}{4} = 1.5 \text{ pounds},$$

as it should.

Example 4. A motorist drives from city A to city B in 2 hours maintaining a constant speed. On the return trip he increases his speed by 5 miles per hour and reduces the travel time to 1 hour 45 minutes. How far apart are A and B, and what were the motorist's speeds?

Solution

Let $d =$ distance from A to B in miles,
and let $x =$ motorist's speed in miles per hour for the initial trip.
Then $x + 5 =$ motorist's speed in miles per hour for the return trip.

Since distance = speed × time,

$$d = 2x$$

and

$$d = (1\tfrac{3}{4})(x + 5).$$

Equating these two expressions for d, one finds

$$2x = \frac{7}{4}x + \frac{35}{4}.$$

Thus

$$8x = 7x + 35$$

$$\therefore x = 35.$$

So the motorist's speeds were 35 mph going and $x + 5 = 40$ mph returning, and cities A and B are $d = 2x = 70$ miles apart.

Check these figures against the original question.

PROBLEMS

1. Rework Problem 1 of Section 4.1 using two unknowns.
2. Rework Problem 8 of Section 4.1 using two unknowns.
3. In 6 years a boy will be three-fourths as old as his sister. Four years ago he was one-half as old as she was. Find their ages.
4. A total of $3000 is invested, part at 6% and part at 7% interest per year. If the total annual interest is $200, how much is invested at each rate?
5. A certain drug is prescribed in the form of a 5% solution. How much of a 2% solution and a 10% solution of this drug must the pharmacist mix to produce 10 ounces of the prescription?
6. An airplane made a trip of 200 miles against the wind in 1 hour. Returning with the wind, the trip took only 50 minutes. Find the speed of the plane in calm air and the speed of the wind.
7. A boat travels 20 miles downstream in 2 hours, but the return trip takes 4 hours. Find the speed of the boat in still water and the speed of the current.
8. A student has taken 10 courses earning grades of A (worth 4 grade points each), B (worth 3 grade points each), and C (worth 2 grade points each). Her grade point average is 3.2, and if the courses in which she got Cs were deleted her average on the remaining courses would be $3.\bar{3}$. How many As, Bs, and Cs did she get? (You can use either two or three unknowns in this problem.)

4.3. Graphing

Numerical data or mathematical relations are often represented by graphs to enhance understanding of the data or relations. Graphical representations can also offer interpretations of the solution of equations.

One begins with the representation of numbers on a straight line. A line is drawn horizontally, as in Figure 1, and equally spaced points are labeled for the integers—increasing to the right. Then each real number corresponds to a point on this line (imagined to extend indefinitely to the right and left). For example, $-5/2$ is located halfway between -3 and -2. The number $\sqrt{2}$ lies a little to the left of the midpoint between 1 and 2.

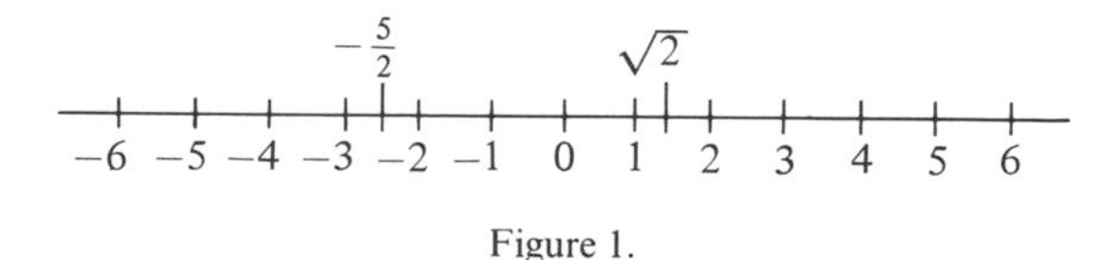

Figure 1.

Note that if a and b are any two numbers with $b > a$, then b is to the right of a on the "number line."

More-interesting ideas can be represented graphically with the aid of two number lines at right angles to each other. Usually, one is drawn horizontally and the other vertically, with their intersection at the zero points on both lines. The horizontal line is usually called the ***x*-axis** and the vertical line, labeled with positive values above zero and negative below, is called the ***y*-axis**. Their point of intersection is called the **origin**. The scales need not be the same on both axes, although they are on the graphs appearing below.

Such a pair of axes now permits one to mathematically describe or label all points in the plane. For example, the point at $x = 1$ and $y = 2$ is reached by starting at the origin and moving right to $x = 1$ on the x-axis, and then moving up 2 units. This point is labeled by the "ordered pair" of numbers (1,2). This means 1 unit to the right (from the origin) and 2 units up. See Figure 2.

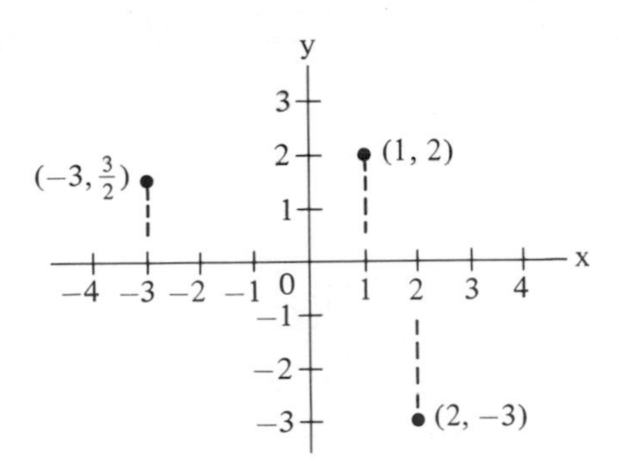

Figure 2.

Similarly, $(2, -3)$ is the point reached by starting at the origin and going 2 units to the right and 3 units *down*. The point $(-3, 3/2)$ is 3 units to the left and $1\frac{1}{2}$ units up.

The values of x and y which locate a point in the plane are referred to as the **coordinates** of that point. Thus $(1, 2)$ is the pair of coordinates for a point, and $(2, -3)$ and $(-3, \frac{3}{2})$ are the pairs of coordinates for two other points.

For convenience in locating points via their coordinates, one sometimes draws a grid of lines parallel to the x- and y-axes as in Figure 3.

The illustrations thus far have just involved locating a few unrelated points in a plane, such as the three points located and labeled in Figure 2. Interesting applications often require graphing infinitely many points on the same paper, with the coordinate pairs all satisfying some common relation, as in the next example.

Example 1. Represent graphically the points with coordinates (x, y) where $y = 2x - 3$.

Solution. Choose various convenient values of x and determine the corresponding values of y from $y = 2x - 3$. The calculations are straightforward, so let us simply summarize the results in the following table:

x	y
-1	-5
0	-3
1	-1
2	1
3	3

Now locate the five points $(-1, -5)$, $(0, -3)$, $(1, -1)$, $(2, 1)$, and $(3, 3)$ in Figure 3. (These points are marked with dots.) These five points *and all others* with coordinates satisfying $y = 2x - 3$ lie on a straight line as shown. (Convince yourself of this by locating a couple more points.)

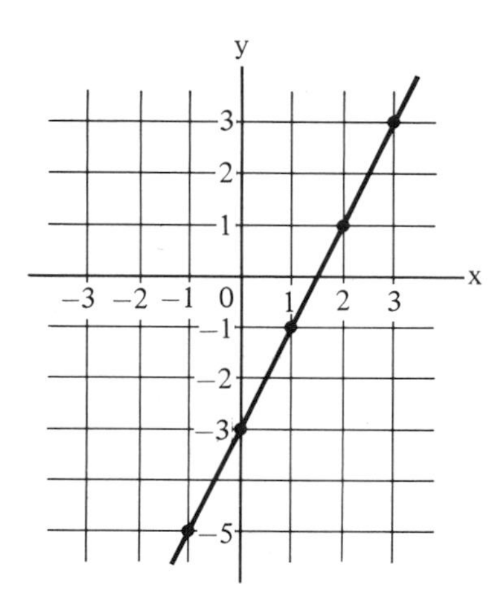

Figure 3.

This straight line is a graphical representation of the relation or equation $y = 2x - 3$.

More generally, any relation of the form

$$y = ax + b,$$

where a and b are constants, has a straight line graph. (You should become convinced of this as you encounter more examples and problems.) Such relations are said to be **linear**.

The problem of solving a system of two equations in two unknowns, as in Section 4.2, has a useful graphical interpretation. This is illustrated by the next example.

Example 2. Represent the equations (or relations)

$$y = 2x - 3 \qquad \text{and} \qquad 2y + x = 2$$

on a single graph. Then use this graph to interpret the concept of "simultaneous solution" of the pair of equations

$$y = 2x - 3$$
$$2y + x = 2.$$

Solution. The information about $y = 2x - 3$ from Example 1 can be used. The second equation can be rewritten as

$$y = -\tfrac{1}{2}x + 1.$$

From this, one can readily compute values of y corresponding to various choices of x. If we use the same values for x as in Example 1, the second equation gives the table

x	y
-1	$3/2$
0	1
1	$1/2$
2	0
3	$-1/2$

This information puts a second straight line on the graph. See Figure 4.

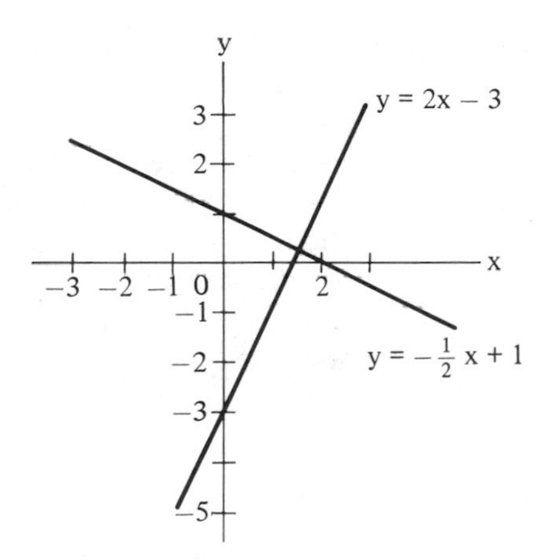

Figure 4.

One point on this graph is special—the point where the two lines cross. It appears to be *approximately* $(\frac{3}{2}, \frac{1}{4})$. That point lies on the graphs of both equations simultaneously. So its coordinates (x, y) must satisfy both the equations

$$y = 2x - 3 \qquad \text{and} \qquad 2y + x = 2$$

simultaneously.

As a check, let us solve these equations by the method of Section 4.2. Rewrite the two equations as

$$y - 2x = -3,$$
$$4y + 2x = \quad 4.$$

Then add to find

$$5y = 1$$
$$\therefore y = \tfrac{1}{5}.$$

Substitute this into, say, the second original equation to get

$$\tfrac{2}{5} + x = 2$$
$$\therefore x = \tfrac{8}{5}.$$

The point with coordinates $(\frac{8}{5}, \frac{1}{5})$ is the exact point at which the two lines cross, and indeed this appears to be confirmed by Figure 4.

(Note that it would not have been possible to obtain the exact values of $x = \frac{8}{5}$ and $y = \frac{1}{5}$, with complete confidence, solely by examining Figure 4. However, if the figure were drawn very carefully, one should be able to come quite close to the exact coordinates by measurements from the graph.)

This example suggests a general interpretation for most of the examples and problems in Section 4.2. Whenever one solves a pair of "linear" equations in two unknowns, one is finding the point of intersection of the two straight lines which form the graphs of these equations (or relations).

In general, two straight lines have exactly one common point (point of intersection). But two other exceptional situations are also possible. (1) The two lines might actually be one and the same line, in which case they have infinitely many points in common. (2) The two lines might be distinct but parallel, in which case they have no points in common.

Example 3. The relations

$$y = 2x - 3 \qquad \text{and} \qquad 2y - 4x = -6$$

represent one and the same line (Figure 5). They have infinitely many points in common.

Example 4. The relations

$$y = 2x - 3 \qquad \text{and} \qquad y - 2x = -2$$

represent two parallel lines (Figure 6). They have no points in common.

Since two points determine a straight line, a *linear* relation can be completely described by the specification of just two of its points.

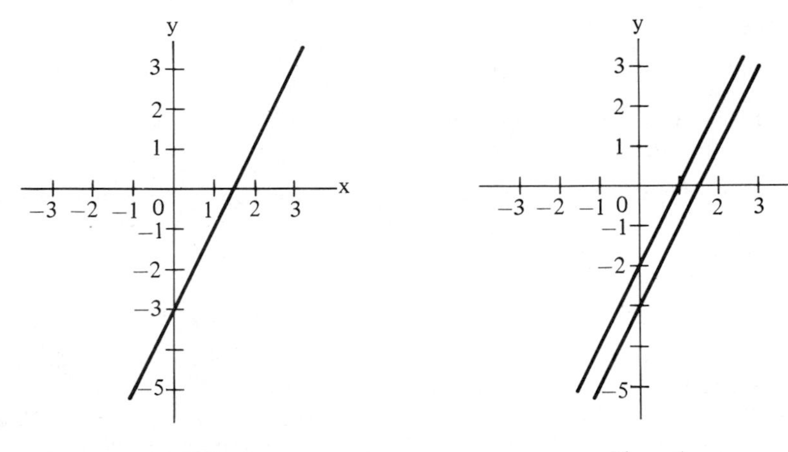

Figure 5.

Figure 6.

Example 5. A publisher determines that he can produce 1000 copies of a certain book for \$10,000 or 4000 copies for \$16,000. Assuming the relation between total cost and number of books produced is linear, find (a) that relation, (b) the cost of producing only 500 books, and (c) the cost of producing 5000 books.

Solution. Let x be the number of books produced and let y be their total cost in dollars. Then, since a linear relation is assumed,

$$y = ax + b$$

for some appropriate choice of the constants a and b. Substitute into this general relation the specific information $x = 1000$, $y = 10{,}000$ to find

$$10{,}000 = 1000a + b.$$

Then substitute $x = 4000$, $y = 16{,}000$ into the general expression to find

$$16{,}000 = 4000a + b.$$

These last two relations are simultaneous equations for the *unknowns* a and b. Subtracting, one finds

$$6000 = 3000a \qquad \therefore\ a = 2.$$

Substituting this value of a back into, say, $10{,}000 = 1000a + b$, one finds $b = 8000$. So the relation between x and y is

$$y = 2x + 8000.$$

Now it is easy to determine that when $x = 500$, $y = \$9000$, and when $x = 5000$, $y = \$18{,}000$.

Problems

1. Graph the relation $3x + 2y = 5$.

2. Graph the relation $y = 2x + 8000$ and confirm from the graph the answers found for questions (b) and (c) of Example 5.

3. In each of the following, attempt to solve the pair of equations for x and y. Then, also graph the relations and compare results.

(a)	(b)	(c)	(d)
$x = 2y - 3$	$x - 40 = 10y$	$x - 2y = 3$	$2x + y = 2$
$2x + 2y = 5$	$5y - 2x - 1 = 0$	$y = \frac{1}{2}x - 1$	$x = -\frac{1}{2}y + 1$

4. The relation between the temperature y in degrees Fahrenheit and x in degrees Celsius is linear. Given that $x = 0°C$ is equivalent to $y = 32°F$, and $x = 100°C$ is equivalent to $y = 212°F$, find and graph the relation between x and y. Now determine the value of y when (a) $x = -40$, (b) $x = 20$, and (c) $x = 40$.

5. The U.S. income tax rate schedule for a single taxpayer for 1983 is shown below. Let $x =$ the amount on form 1040, line 37, the "taxable income" in dollars, and let

Schedule X
Single Taxpayers

Use this Schedule if you checked **Filing Status Box 1** on Form 1040—

If the amount on Form 1040, line 37 is: Over—	But not over—	Enter on Form 1040, line 38	of the amount over—
$0	$2,300	**—0—**	
2,300	3,400	**.......11%**	**$2,300**
3,400	4,400	**$121 + 13%**	**3,400**
4,400	8,500	**251 + 15%**	**4,400**
8,500	10,800	**866 + 17%**	**8,500**
10,800	12,900	**1,257 + 19%**	**10,800**
12,900	15,000	**1,656 + 21%**	**12,900**
15,000	18,200	**2,097 + 24%**	**15,000**
18,200	23,500	**2,865 + 28%**	**18,200**
23,500	28,800	**4,349 + 32%**	**23,500**
28,800	34,100	**6,045 + 36%**	**28,800**
34,100	41,500	**7,953 + 40%**	**34,100**
41,500	55,300	**10,913 + 45%**	**41,500**
55,300	....	**17,123 + 50%**	**55,300**

$y =$ the tax in dollars found from this schedule (to be entered on Form 1040, line 38). (a) Write out the relation between x and y assuming $x \geq 55{,}300$. (b) Graph this relation. (c) Determine the tax on a "taxable income" of \$100,000.

6. (a) Find the relation between x and y as defined in Problem 5 for a single taxpayer with "taxable income" between \$18,200 and \$23,500. (b) Graph this relation on the same sheet with the graph of Problem 5. (c) Determine the tax on a "taxable income" of \$20,000.

7. If it is 50°F in Minneapolis and 100°F in Dallas, does it make sense to say that Dallas is twice as hot as Minneapolis?

8. Which is colder: (a) −20°F or −20°C? (b) −50°F or −50°C?

CHAPTER 5

Quadratic Polynomials and Equations

The equation $2x + 4 = 7$ is called a "linear" equation. This type of problem arose repeatedly in Chapter 4.

A slightly more complicated equation would be

$$2x^2 + 5x + 4 = 7.$$

This is an example of a "quadratic" equation. Such equations, or just quadratic polynomials (or quadratic relations) such as $y = 2x^2 + 5x + 4$, arise in certain time–speed–distance problems and other applications to be considered in Section 5.2.

First comes a review of the methods of solving quadratic equations with special emphasis on a process called "completing the square."

5.1. Solution of Quadratic Equations

The most general **quadratic equation** to be considered can be written in "standard form" as

$$ax^2 + bx + c = 0,$$

where a, b, c stand for given numbers (with a $\neq 0$), and x is the unknown. The presence of the x^2 term distinguishes this from a "linear" equation.

Again, one usually wants to solve for the unknown x.

Example 1. A particularly simple quadratic equation is $x^2 - 4 = 0$. In this case, one can add 4 to both sides, getting $x^2 = 4$. Then observe that there are two solutions, $x = 2$ and $x = -2$.

These two answers are sometimes written more compactly as $x = \pm 2$, pronounced "x equals plus or minus 2."

This property of having two solutions is typical of quadratic equations. However, the method of solution used in Example 1 is *not* useful for most quadratics. Here is a more general approach.

Example 2. Solve $x^2 - 6x + 4 = 20$.

Solution. Subtract 20 from both sides to put the equation in standard form

$$x^2 - 6x - 16 = 0.$$

Now attempt to factor the left-hand side. If you are out of practice with factoring, this may be difficult. However, if someone simply tells you that

$$x^2 - 6x - 16 = (x - 8)(x + 2),$$

you should be able to *verify* it by multiplication.

It follows that the original quadratic equation is equivalent to

$$(x - 8)(x + 2) = 0.$$

Now, one property of the real numbers is that a product is zero if and only if at least one of the factors is zero. Thus, one concludes that either

$$x - 8 = 0 \qquad \text{or} \qquad x + 2 = 0.$$

Therefore

$$x = 8 \qquad \text{or} \qquad x = -2.$$

These values *should be checked by substitution* into the original equation. Putting $x = 8$ into the left-hand side, one finds

$$x^2 - 6x + 4 = 8^2 - 6(8) + 4 = 64 - 48 + 4$$

which equals 20, as it should. Similarly, $x = -2$ gives

$$x^2 - 6x + 4 = (-2)^2 - 6(-2) + 4 = 4 + 12 + 4 = 20.$$

The check by substitution is the ideal method for confirming the accuracy of your work.

In Example 2, the factorization of $x^2 - 6x - 16$ was produced without explanation. The next example illustrates an important method—called "completing the square"—for systematically finding the factorization in very general cases.

Example 3 (Completing the Square). Solve $2x^2 + 5x + 4 = 7$.

Solution. Subtract 7 from both sides to put the equation in standard form,

$$2x^2 + 5x - 3 = 0.$$

Now the factorization of the left-hand side is probably *not* obvious.

First divide through by the coefficient of x^2, which is 2 in this case. Thus an equivalent quadratic equation is

$$x^2 + \frac{5}{2}x - \frac{3}{2} = 0.$$

To "complete the square," one examines the terms involving x, namely, $x^2 + \frac{5}{2}x$. Imagine these as the x terms which would result from squaring the expression $x + d$, for some appropriate choice of d. Since

$$(x + d)^2 = x^2 + 2dx + d^2,$$

in order to obtain agreement with $x^2 + \frac{5}{2}x$, one must set

$$2d = \tfrac{5}{2} \qquad \text{or} \qquad d = \tfrac{5}{4}.$$

Then

$$(x + d)^2 = \left(x + \frac{5}{4}\right)^2 = x^2 + \frac{5}{2}x + \frac{25}{16}.$$

Now rewrite the quadratic equation $x^2 + \frac{5}{2}x - \frac{3}{2} = 0$ with the number 25/16 *both added and subtracted* on the left-hand side. This gives

$$x^2 + \frac{5}{2}x + \frac{25}{16} - \frac{25}{16} - \frac{3}{2} = 0,$$

which is *equivalent* to the original equation. The addition *and* subtraction of 25/16 makes no net change. But now the equation can be recognized as

$$\left(x^2 + \frac{5}{2}x + \frac{25}{16}\right) - \frac{49}{16} = 0$$

or

$$(x + \tfrac{5}{4})^2 - (\tfrac{7}{4})^2 = 0.$$

The left-hand side of this can be factored if one recalls the "difference of two squares"

$$y^2 - z^2 = (y - z)(y + z).$$

Check this!

Set $y = x + \frac{5}{4}$ and $z = \frac{7}{4}$. Then

$$(x + \tfrac{5}{4})^2 - (\tfrac{7}{4})^2 = (x + \tfrac{5}{4} - \tfrac{7}{4})(x + \tfrac{5}{4} + \tfrac{7}{4}).$$

So the equation becomes

$$(x + \tfrac{5}{4} - \tfrac{7}{4})(x + \tfrac{5}{4} + \tfrac{7}{4}) = 0$$

or

$$(x - \tfrac{1}{2})(x + 3) = 0.$$

Once again, the product of two numbers is zero if and only if (at least) one of the factors is zero. Thus either

$$x - \tfrac{1}{2} = 0 \qquad \text{or} \qquad x + 3 = 0.$$

In other words, either

$$x = \tfrac{1}{2} \qquad \text{or} \qquad x = -3.$$

Again, these values should be checked by substitution into the original equation, $2x^2 + 5x + 4 = 7$. Do it and this example will be complete.

Factorization, by completing the square if necessary, is the general method for solving quadratic equations. In fact, it can be used to derive the following solution to the general quadratic equation.

The Quadratic Formula. *The equation*

$$ax^2 + bx + c = 0,$$

where $a \neq 0$, b, and c are given constants, has the solutions

$$x = \frac{-b \pm \sqrt{b^2 - 4ac}}{2a}.$$

(Using first the + sign and then the − sign, one gets two different solutions, except when $b^2 - 4ac = 0$.)

PROOF. The following steps, including completing the square, correspond to those in Example 3.

First divide the original equation through by a. This gives

$$x^2 + \frac{b}{a}x + \frac{c}{a} = 0.$$

Now choose the number d so that when one computes

$$(x + d)^2 = x^2 + 2dx + d^2,$$

the first two terms on the right-hand side agree with $x^2 + (b/a)x$. This requires

$$2d = \frac{b}{a},$$

which means

$$d = \frac{b}{2a}.$$

Then

$$\begin{aligned}\left(x + \frac{b}{2a}\right)^2 &= x^2 + 2\left(\frac{b}{2a}\right)x + \left(\frac{b}{2a}\right)^2 \\ &= x^2 + \frac{b}{a}x + \frac{b^2}{4a^2}.\end{aligned}$$

Seeing this, one goes back to the quadratic equation $x^2 + (b/a)x + c/a = 0$. Add and subtract $b^2/(4a^2)$ on the left-hand side to get

$$x^2 + \frac{b}{a}x + \frac{b^2}{4a^2} - \frac{b^2}{4a^2} + \frac{c}{a} = 0$$

or

$$\left(x + \frac{b}{2a}\right)^2 - \frac{b^2 - 4ac}{4a^2} = 0.$$

In other words,

$$\left(x + \frac{b}{2a}\right)^2 - \left(\frac{\sqrt{b^2 - 4ac}}{2a}\right)^2 = 0.$$

Now, recognizing this as the difference of two squares, one gets the factored form

$$\left(x + \frac{b}{2a} - \frac{\sqrt{b^2 - 4ac}}{2a}\right)\left(x + \frac{b}{2a} + \frac{\sqrt{b^2 - 4ac}}{2a}\right) = 0.$$

So again it follows that (at least) one of these factors must be zero. Thus either

$$x = \frac{-b + \sqrt{b^2 - 4ac}}{2a} \quad \text{or} \quad x = \frac{-b - \sqrt{b^2 - 4ac}}{2a}.$$

This is the assertion of the quadratic formula. □

In some quadratic equations $b^2 - 4ac$ is negative, and there is *no real number* which represents $\sqrt{b^2 - 4ac}$ in this case. Why? (The solutions of the quadratic equation can then be described in terms of "imaginary" or "complex" numbers.) In this book, any quadratic equation for which $b^2 - 4ac < 0$ will simply be said to have *no real solutions*.

Example 4. Solve $2x^2 + 5x + 4 = 7$, the equation of Example 3, with the aid of the quadratic formula.

Solution. Subtract 7 from both sides to put the equation into standard form:

$$2x^2 + 5x - 3 = 0.$$

Thus $a = 2$, $b = 5$, and $c = -3$. So the solutions are

$$x = \frac{-b \pm \sqrt{b^2 - 4ac}}{2a} = \frac{-5 \pm \sqrt{25 + 24}}{4} = \frac{-5 \pm 7}{4} = \frac{1}{2} \text{ and } -3.$$

Again, the solutions should be checked by substitution into the original quadratic equation.

Example 5. Solve $4x^2 = 4x + 7$ by means of the quadratic formula.

Solution. Subtract $4x + 7$ from both sides to get the standard form

$$4x^2 - 4x - 7 = 0.$$

Thus $a = 4$, $b = -4$, $c = -7$, and the solutions are

$$x = \frac{-b \pm \sqrt{b^2 - 4ac}}{2a} = \frac{4 \pm \sqrt{16 + 112}}{8} = \frac{4 \pm \sqrt{64 \times 2}}{8}$$

$$= \frac{1}{2} \pm \sqrt{2}.$$

If, for any reason, you needed to convert these two solutions to decimal form, you could use $\sqrt{2} \simeq 1.414$ and find

$$x = 0.5 + \sqrt{2} = 1.914$$

or

$$x = 0.5 - \sqrt{2} = -0.914.$$

PROBLEMS

1. Solve the following quadratic equations by factoring, *without using the quadratic formula*. (When necessary, do use the technique of completing the square.) Check each answer by substitution.
   (a) $x^2 + 3x = 4$
   (b) $x^2 + 25 = 10x$
   (c) $x^2 - 7 = 0$
   (d) $3x = 2(x^2 - 1)$
   (e) $6x^2 - 5 = 13x$
   (f) $x^2 + 2x - 2 = 0$
   (g) $x^2 + 2x + 2 = 0$.
2. Solve the same equations as in Problem 1 by means of the quadratic formula.

5.2. Applications of Quadratic Equations

The examples and problems of this section describe some situations which might naturally give rise to quadratic equations.

Example 1. A motorboat travels at a constant speed of 9 mph in still water. If this boat travels 1 mile upstream and immediately returns, requiring 15 minutes for the round trip, find the speed of the current.

Solution

Let $x =$ the speed of the current in miles per hour.

Then the boat travels upstream at $9 - x$ mph, and downstream at $9 + x$ mph. So the trip of 1 mile in each direction takes a total time of

$$\frac{1}{9 - x} + \frac{1}{9 + x} = \frac{1}{4} \text{ hour.}$$

Now multiply both sides of this equation by $4(9 - x)(9 + x)$ to find

$$4(9 + x) + 4(9 - x) = (9 - x)(9 + x)$$

or

$$72 = 81 - x^2.$$

So $x^2 = 9$, or $x = \pm 3$.

The desired answer is the positive solution, $x = 3$ mph.

The negative solution, $x = -3$, will simply be discarded as "extraneous." (Actually, in this case, it could be interpreted as a possible current speed in the opposite direction.)

Example 2. An investment club decided to make a certain stock purchase for a total of \$9000, with each member paying an equal share of this price. But, before the transaction was completed, two members left the club and the remaining members had to pay an extra \$50 each in order to still produce the \$9000. How many members were in the club initially?

Solution

Let $x =$ the number of members in the club initially.

Then the original cost per member was \$9000/x. After two members left the club, the cost per remaining member became \$9000/x + 50, which had to equal \$9000/(x − 2). Thus

$$\frac{9000}{x} + 50 = \frac{9000}{x - 2}.$$

Multiply both sides of this equation by $x(x - 2)$ to find

$$9000(x - 2) + 50x(x - 2) = 9000x$$

or

$$50x^2 - 100x - 18{,}000 = 0.$$

The ensuing arithmetic will be simplified slightly if one first divides the equation through by 50 to obtain

$$x^2 - 2x - 360 = 0.$$

If you are clever enough (or lucky enough) to factor the left-hand side you will obtain

$$(x - 20)(x + 18) = 0.$$

Thus, either $x - 20 = 0$ or $x + 18 = 0$. This leads to solutions

$$x = 20 \qquad \text{or} \qquad x = -18.$$

(Or these solutions can be obtained by completing the square or using the quadratic formula.)

The negative answer is rejected at once as "extraneous." The remaining solution, $x = 20$, should be checked by substitution.

Quadratic equations also arise in the study of an object in "free fall."

Free Fall. An object is said to be **in free fall** or **falling freely** if it is moving solely under the influence of gravity. (It may actually be going upward if it was initially thrown upward and has not yet started descending.)

In the discussion of free fall motion, we shall ignore the effects of air resistance and of the rotation of the earth, and we shall assume that the heights involved are very small compared to the radius of the earth.

Under these assumptions, it is known that an object projected upward with an initial velocity v_0 feet/second from an initial height h_0 feet above the surface of the earth will be at (approximately) height

$$h = h_0 + v_0 t - 16t^2 \text{ feet after } t \text{ seconds.}$$

In the above, v_0 is positive if the object is initially thrown or fired upwards, v_0 is negative if the object is initially fired downwards, and $v_0 = 0$ if the object is simply dropped. Similarly, h and h_0 represent heights above the surface of the earth if they are positive. They represent depths below the surface (as in a well) if they are negative.

Example 3. A baseball is thrown straight upwards from ground level ($h_0 = 0$) with an initial velocity of 50 feet per second. When (if ever) will it reach a height of 30 feet?

Solution. Since $h_0 = 0$ and $v_0 = 50$, the height after t seconds will be

$$h = 50t - 16t^2 \text{ feet.}$$

Set this equal to 30 feet, and try to solve for t. Thus

$$30 = 50t - 16t^2$$

or (in standard form)

$$16t^2 - 50t + 30 = 0.$$

Divide through by 2 in order to work with smaller numbers. Then

$$8t^2 - 25t + 15 = 0$$

and the quadratic formula gives

$$t = \frac{25 \pm \sqrt{25^2 - 4(8)(15)}}{16} = \frac{25 \pm \sqrt{625 - 480}}{16}$$

$$= \frac{25 \pm \sqrt{145}}{16}.$$

To compute $\sqrt{145}$, use the method of Section 3.3. Starting with a first guess of $x_1 = 12$, one quickly finds $x_2 = (145/12 + 12) \div 2 = 12.04$, which is a good approximation to $\sqrt{145}$. Thus

$$t = \frac{25 \pm 12.04}{16} = 0.81 \qquad \text{or} \qquad 2.315 \text{ seconds.}$$

In this example, both answers make sense. The first indicates when the baseball reaches a height of 30 feet on the way up, and the second shows when it passes this level again on the way down.

Problems

1. Find two consecutive integers whose product is 272.
2. Find two consecutive odd numbers whose product is 323.
3. A rectangular field is 50 feet longer than it is wide, and its area is 30,000 square feet. Find its dimensions.
4. One leg of a right triangle is 7 cm longer than the other, and the hypotenuse is 17 cm long. Find the lengths of the two legs.
5. A plane flying with a constant air speed of 225 mph makes a trip of 1000 miles against the wind, and returns with the wind in a total of 9 hours. Assuming the wind speed is constant, find the wind speed.
6. If a number equals its own square, what is it?
7. A group of x persons charters a block of x seats on a plane flight for a total of \$3600. When three members of the group drop out and fail to pay their share, the remaining members have to pay an additional \$40 each to make up the total of \$3600. How many were in the group originally?
8. A stone is dropped into a mine shaft 800 feet deep. How long does it take to reach the bottom?
9. A skyrocket is launched from the ground, and at the instant it stops burning it is 50 feet above the ground and headed straight upward at a speed of 80 feet per second. (From this instant onward it is a problem of "free fall.") When will the rocket reach a height of (a) 130 feet, (b) 150 feet, (c) 170 feet?
10. How long after the rocket in Problem 9 stops burning will it hit the ground?
11. When does the baseball in Example 3 reach a height of 50 feet?
12. If the length of each edge of a certain cube were increased by 1 inch, the volume of the cube would increase by 91 cubic inches. What is the length of an edge?

5.3. Quadratic Polynomials

An expression of the form

$$ax^2 + bx + c,$$

where $a \neq 0$, b, and c are given numbers, is called a **quadratic polynomial** in x.

The previous two sections were primarily aimed at solving quadratic *equations* such as $ax^2 + bx + c = 0$. Now consider the more general problem of finding the values of $ax^2 + bx + c$ for various values of x, and graphing the relation

$$y = ax^2 + bx + c.$$

Example 1. Graph the simplest quadratic polynomial relation,

$$y = x^2.$$

Solution. As in Section 4.3, one begins by listing several values of x together with the corresponding values of y:

x	y
-3	9
-2	4
-1	1
0	0
1	1
2	4
3	9

This table gives the coordinate pairs for seven points on the graph, and these are displayed in Figure 1. (The x and y scales are chosen unequal here for convenience.) The graph drawn through the seven points is a *smooth* curve. It is not a straight line as in Section 4.3. Note that the graph is symmetric in the sense that two values of x which differ only in algebraic sign give the same value of y. For example, $x = 2$ and $x = -2$; both give $y = 4$.

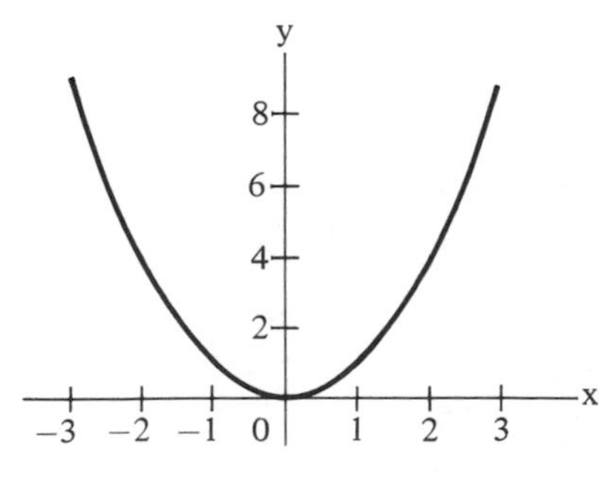

Figure 1.

Note also that there is no sharp point at $(0, 0)$. To convince yourself of this, calculate and locate on the graph the points corresponding to $x = \pm\frac{1}{2}$ and then $x = \pm\frac{1}{4}$. See Problem 1.

This graph is "representative" of many quadratic polynomials. In fact, the graph of the relation

$$y = ax^2 + bx + c$$

will have the same *general* shape as the graph in Figure 1 whenever $a > 0$. However, it will not, in general, be located in the same place relative to the x- and y-axes, and it will not usually have the same proportions.

Before calculating the required table of coordinate pairs for such a graph, it is often convenient to rewrite the relation by "completing the square." This is essentially the procedure used in solving quadratic equations in Section 5.1. But now, instead of dividing through by a, merely "factor out" a from the x^2 and x terms; then *keep* the multiplier a as part of the definition of y.

Thus one writes

$$\begin{aligned} y &= ax^2 + bx + c \\ &= a\left(x^2 + \frac{b}{a}x \qquad \right) \qquad + c. \end{aligned}$$

Now, as before, one wants to produce a quantity inside the parentheses of the form $x^2 + 2dx + d^2$. This requires $2d = b/a$, or $d = b/(2a)$. So add $d^2 = b^2/(4a^2)$ inside the parentheses. This addition is compensated outside the parentheses by subtracting ad^2. Can you see why?

Example 2. Complete the square and graph the relation

$$y = 2x^2 + 4x - 3.$$

Solution. Rewrite the relation as

$$\begin{aligned} y &= 2x^2 + 4x - 3 \\ &= 2(x^2 + 2x \qquad) \qquad - 3 \\ &= 2(x^2 + 2x + 1) - 2 - 3. \end{aligned}$$

Note that the addition of 1 inside the parentheses is compensated by the subtraction of 2 outside. Thus

$$y = 2(x + 1)^2 - 5.$$

With the relation written in this form, one can show that the minimum possible value of y (namely, $y = -5$) must occur when $x = -1$. For then $2(x + 1)^2 = 0$, while any other value of x, greater or smaller, will make $2(x + 1)^2 > 0$ and hence $y > -5$. Moreover, the graph will be symmetrical about the value $x = -1$. For example, $x = -2$ and $x = 0$ (one unit to the left of -1 and one unit to the right of -1) both give the same value of y, namely, $2(1)^2 - 5 = -3$.

Some convenient coordinate pairs, computed from $y = 2(x + 1)^2 - 5$, are listed and a graph of the relation is sketched in Figure 2. Here, one clearly sees the special nature of the point $(-1, -5)$ where y takes its minimum value.

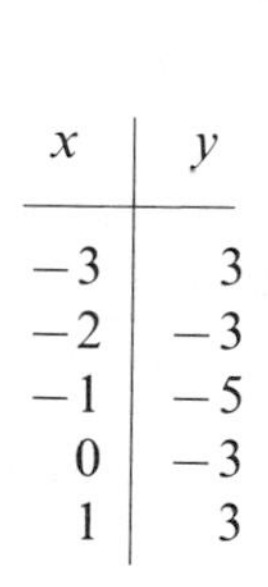

x	y
-3	3
-2	-3
-1	-5
0	-3
1	3

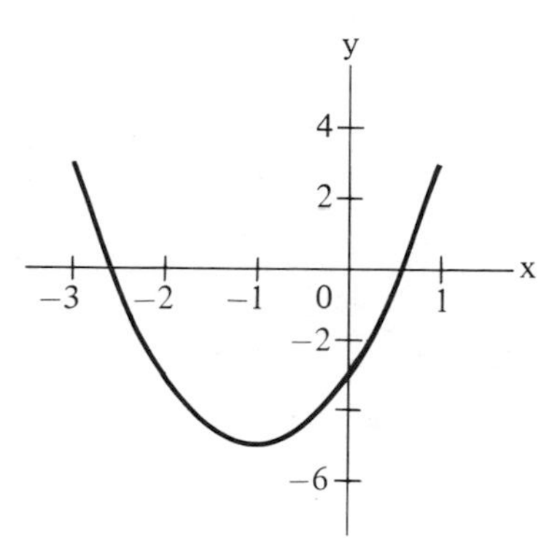

Figure 2.

In case $a < 0$, the procedure for graphing the relation

$$y = ax^2 + bx + c$$

is similar. But now, as you will see the curve is inverted and y achieves a maximum value rather than a minimum.

Example 3. Graph the relation

$$y = -2x^2 + 5x + 4.$$

Solution. Again factor a out of the x^2 and x terms before completing the square. Thus write

$$\begin{aligned} y &= -2x^2 + 5x + 4 \\ &= -2(x^2 - \tfrac{5}{2}x \qquad\quad) + 4 \end{aligned}$$

To complete the square, one must now add 25/16 inside the parentheses and compensate for this by *adding* $2(25/16) = 25/8$ outside. Thus

$$\begin{aligned} y &= -2\left(x^2 - \frac{5}{2}x + \frac{25}{16}\right) + \frac{25}{8} + 4 \\ &= -2\left(x - \frac{5}{4}\right)^2 + \frac{57}{8}. \end{aligned}$$

Now observe that when $x = 5/4$, $y = 57/8$; and any other value of x makes $-2(x - 5/4)^2$ negative and hence $y < 57/8$. Thus 57/8 is the maximum possible value of y. A table of (x, y) values is computed, and then the resulting graph is given in Figure 3.

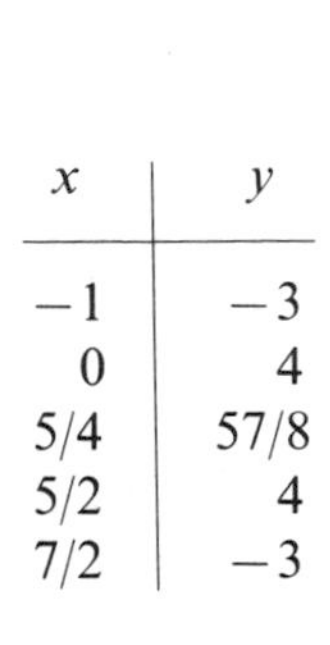

x	y
-1	-3
0	4
$5/4$	$57/8$
$5/2$	4
$7/2$	-3

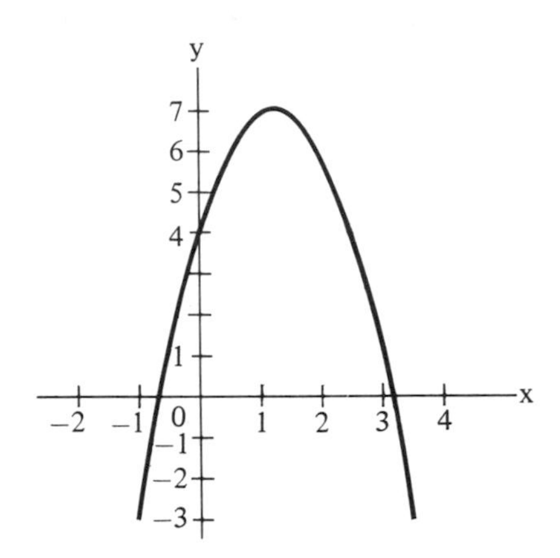

Figure 3.

In the above table, the rather curious values $x = \frac{5}{4}$, $x = \frac{5}{2}$, and $x = \frac{7}{2}$ were each chosen for a reason. The value $x = \frac{5}{4}$ was used, of course, because this produces the maximum value of y. Then $x = \frac{5}{2}$ and $x = \frac{7}{2}$ were used because they are just as far to the right of $x = \frac{5}{4}$ as the values $x = 0$ and $x = -1$ are to the left. Hence, thanks to the symmetry about $x = \frac{5}{4}$, the value of y when $x = \frac{5}{2}$ is the same as the value of y when $x = 0$; and the value of y when $x = \frac{7}{2}$ is the same as the value of y when $x = -1$.

You may very well wonder how one can expect to draw the graph of Figure 3 with any degree of confidence after locating only five points. This is possible only if one has already had some experience with quadratic polynomials (experience possibly gained from Examples 1 and 2), and one knows where the maximum value of y occurs and what kind of symmetry the graph has. For further confirmation that Figure 3 is correct, compute some more points and locate them on the graph. See Problem 2.

Before leaving these examples, note the significance of the points where the graph of $y = ax^2 + bx + c$ crosses the x-axis. These values of x are the *solutions* of the quadratic equation $ax^2 + bx + c = 0$. See Problems 3 and 4.

The final examples of this section offer interesting applications.

Example 4. If you have p meters of fence to be used to enclose a rectangular field, what should be the dimensions of the rectangle in order to enclose the greatest possible area?

Solution

Let $x =$ one dimension of the rectangle in meters.

Then, since the perimeter p is the sum of the lengths of all four sides, the other dimension is $p/2 - x$ meters. The area of the rectangle is, therefore,

$$A = \left(\frac{p}{2} - x\right)x = -x^2 + \frac{p}{2}x \text{ square meters.}$$

Here p is a constant which we cannot change, but x can be any number between 0 and $p/2$. Why?

Since A is a quadratic polynomial in x, we seek the maximum possible value of A via completing the square. Thus

$$\begin{aligned} A &= -\left(x^2 - \frac{p}{2}x \qquad \right) \\ &= -\left(x^2 - \frac{p}{2}x + \frac{p^2}{16}\right) + \frac{p^2}{16} \\ &= -\left(x - \frac{p}{4}\right)^2 + \frac{p^2}{16}. \end{aligned}$$

Now, arguing as in previous examples, we see that when $x = p/4$, $A = p^2/16$. Any other choice for x makes $A < p^2/16$.

So, the maximum possible area results when $x = p/4$. And in this case the other dimension of the rectangle is also $p/4$.

It follows that the rectangle of maximum area with a given perimeter is always a square.

Example 5. A baseball is thrown straight upwards from a height of 7 feet with an initial velocity of 50 feet per second. Find the maximum height of the baseball and the time when this height is reached.

Solution. The data is similar to that in Example 3 of Section 5.2. The height after t seconds is

$$h = 7 + 50t - 16t^2 \text{ feet.}$$

Rewrite this to complete the square as follows:

$$\begin{aligned} h &= -16\left(t^2 - \frac{25}{8}t\right) + 7 \\ &= -16\left(t^2 - \frac{25}{8}t + \frac{25^2}{16^2}\right) + 16\left(\frac{25^2}{16^2}\right) + 7 \\ &= -16\left(t - \frac{25}{16}\right)^2 + \frac{625}{16} + 7. \end{aligned}$$

Then one can read off the maximum height of $625/16 + 7 = 46$ feet (rounded off), and observe that it is reached when $t = 25/16$ seconds after the ball is thrown.

Problems

1. Compute the value of $y = x^2$ when $x = \pm\frac{1}{2}$ and when $x = \pm\frac{1}{4}$, and locate the resulting four points on Figure 1.

2. Compute the value of $y = -2x^2 + 5x + 4$ when $x = 1$, when $x = 2$, and when $x = 3$, and locate the resulting three points on Figure 3.

3. (a) Solve the equation $2x^2 + 4x - 3 = 0$, and interpret your solutions on Figure 2.
   (b) Solve the equation $-2x^2 + 5x + 4 = 0$, and interpret on Figure 3.

4. Graph the following three quadratic polynomials on one and the same figure. Pay particular attention to the maximum or minimum value of y for each graph. And, in each case, solve the equation $y = 0$ for x and interpret the result(s) on the graph.
   (a) $y = x^2 + 2x - 3$, (b) $y = x^2 + 2x + 1$, (c) $y = x^2 + 2x + 2$.

5. A stone is thrown straight upwards from a height of 4 feet above the ground with a speed of 48 feet/second.
   (a) Find the maximum height reached by the stone.
   (b) When does the stone hit the ground?

6. You have 500 feet of fence to enclose a rectangular field for grazing cattle. One edge of the field will be alongside a steep cliff, so you need not waste fence on that side. (Did you ever see a cow climbing a cliff?) What dimensions should you use for the rectangle in order to enclose the maximum possible area?

7. Find the maximum height reached by the skyrocket in Problem 9 of Section 5.2 and the instant when this height is reached.

*8. Show that, for every integer n, $n^3 - n$ is divisible by 6. [*Hint*. Factor the "polynomial" $n^3 - n$ and examine the factors.]

CHAPTER 6

Powers and Geometric Sequences

What do pyramid clubs, hot cups of coffee, gambling outcomes, and inflation rates have in common? Consider the following specific problems.

(a) If each recipient of a certain chain letter dutifully delivers copies of the same letter to just five other persons in 1 week, how many people will be involved in the first 8 weeks?
(b) A cup of coffee in a room where the temperature is 70°F cools from 190°F to 130°F in 6 minutes. How much longer will it take to cool down to 85°F?
(c) Assume you have $100 for gambling at the roulette wheel, and at each spin of the wheel you bet half the money you then have on red, which is an "even-money" bet. Thus you win or lose the amount of your bet. If you win 10 times and lose 10 times will you end up ahead or behind or even?
(d) If the Consumer Price Index rises 1% per month for 12 months, what is its rise for the entire year?

The answers to these problems, incidentally, are (a) about half a million people, (b) about 12 *more* minutes, (c) you will have only about $5.63 left (regardless of the order of your wins and loses), and (d) approximately 12.7%.

The unifying feature of these problems—and others on population growth, compound interest, and archeological dating—is the fact that their solutions all involve successive powers of some fixed number. For example, (a) uses the powers $5, 5^2, 5^3, 5^4, \ldots$ and (d) uses the successive powers $1.01, (1.01)^2, (1.01)^3, \ldots$.

One of the beauties of mathematics is that it unifies such diverse problems. Thus, similar ideas are useful in very different settings.

6.1. Applications of Powers

The examples and problems of this section illustrate some simple uses of powers.

Example 1. A certain chain letter asks each recipient to send copies of the same letter to five other persons. Assume that the person who initiated the letter, referred to as "level 1," sends out five copies and these reach their destinations in 1 week. Then each of the five recipients, referred to as "level 2," promptly sends out five copies which also reach their destinations in 1 week; and the chain continues unbroken in this manner. How many people will be involved in the first 8 weeks?

Solution. Within 1 week from the start, the five people in "level 2" will have received their letters. So the total number involved will be $1 + 5 = 6$ (the initiator and the first five recipients). By the end of the second week, the $5^2 = 25$ people in "level 3" will have received their letters and there will be a total of $1 + 5 + 5^2$ people involved. Continuing in this manner, one sees that at the end of 8 weeks all the people through "level 9" will have received letters. The numbers of people added at each level up to and including level 9 are, respectively,

$$1, 5, 5^2, 5^3, 5^4, 5^5, 5^6, 5^7, \text{ and } 5^8,$$

or

$$1, 5, 25, 125, 625, 3125, 15{,}625, 78{,}125, \text{ and } 390{,}625.$$

The total number of people involved is the sum of these numbers, 488,281.

In general, if a and r are two fixed numbers, then

$$a, ar, ar^2, ar^3, ar^4, \ldots$$

is called a **geometric sequence**. Example 1 makes use of the geometric sequence with $a = 1$ and $r = 5$. (Think of r as standing for the common *ratio* between consecutive terms in the sequence.)

Radioactive materials decay at a rate which is always proportional to the present amount of material. Thus if an initial quantity Q_0 of some radioactive material decays to just $\frac{1}{2}Q_0$ in time h, then it will decay to just half of this, namely, $\frac{1}{4}Q_0$ in an *additional* time h. The value of h is called the **half-life** of the material. It is different for different radiocative substances.

Example 2. Radium 226 has a half-life of $h = 1620$ years. If one starts with $Q_0 = 10$ grams of radium 226, how much will remain after 1620 years, 3240 years, 4860 years, etc.?

Solution. Since 1620 years is the half-life, the amount remaining after 1620 years will be $10(\frac{1}{2}) = 5$ grams. After another 1620 years (3240 altogether) only half of this will remain, i.e., $10(\frac{1}{2})^2 = 2.5$ grams. Thus, starting with the initial 10 grams, and proceeding by 1620-year intervals, the amounts are

$$10, \quad 10(\tfrac{1}{2}), \quad 10(\tfrac{1}{2})^2, \quad 10(\tfrac{1}{2})^3, \ldots.$$

This is a geometric sequence with $a = 10$ and $r = \frac{1}{2}$.

If one lets Q be the quantity of radium 226 remaining after t years, then $Q = Q_0$ when $t = 0$, $Q = Q_0(\frac{1}{2})$ when $t = 1620$, $Q = Q_0(\frac{1}{2})^2$ when $t = 3240$, ... In this usage, Q represents *different* numbers for different values of t.

Newton's Law of Cooling describes the changing temperature of a heated object which is left to cool in an environment having a *constant* temperature. Specifically, consider the *difference* $T - T_a$, where T is the temperature of the heated object and T_a is the constant temperature of its surroundings (the ambient temperature). Then $T - T_a$ will decrease at a rate which is always proportional to its present value.

Notice how similar this is to the sentence describing radioactive decay on page 78. Thus the half-life concept now applies to $T - T_a$.

Specifically, if the object has an initial temperature T_0, then after a certain time h the temperature T will satisfy

$$T - T_a = (T_0 - T_a)(\tfrac{1}{2}).$$

The time h required to achieve this will be called the half-life.

Example 3. A pan of water is heated to boiling point, 100°C, and then is removed to cool in a room with constant temperature $T_a = 20$°C. If the pan cools to 40°C in 50 minutes, when will it reach 30°C?

Solution. Let $T =$ the temperature of the pan in degrees Celsius t minutes after removal from the stove. (Thus *different* values of T are associated with different values of t.)

The statement of the problem provides the information that $T = T_0 = 100$ when $t = 0$. Next, one would like to determine the half-life (in minutes). The temperature *difference* $T - T_a$, or $T - 20$ in this case, is reduced to half its former value every h minutes. Since the initial value of $T - T_a$ is $100 - 20 = 80$, the values of $T - T_a$ at the ends of successive h-minute intervals will be

$$80, \quad 80(\tfrac{1}{2}), \quad 80(\tfrac{1}{2})^2, \quad 80(\tfrac{1}{2})^3, \quad 80(\tfrac{1}{2})^4, \quad \ldots$$

or

$$80, \quad 40, \quad 20, \quad 10, \quad 5, \quad \ldots.$$

Now, from the given information, when $t = 50$ one has $T - T_a = 20$. Referring to the above sequence of values of $T - T_a$, one notes that

$T - T_a = 20$ when $t = 2h$. So $2h = 50$ or $h = 25$. (The half-life is 25 minutes.)

The question "when will $T = 30$?" is equivalent to the question "when will $T - T_a = 10$?" Another glance at the sequence of values of $T - T_a$ shows that this occurs when $t = 3h$, i.e., 75 minutes after removal from the stove.

Problems

1. Imagine a large piece of paper five-thousandths of an inch thick being torn in half, and the two pieces placed one on top of the other. Then these two pieces are torn in half, and the four pieces are put in a stack. This process of tearing in half and stacking is done 25 times. Will the final stack be (a) less than an inch thick, (b) between an inch and a yard thick, or (c) more than a yard thick?
2. The radioactive isotope carbon 14 has a half-life of about 5700 years. From an initial quantity of 1 gram in the year 15,000 B.C., how much would remain today (approximately)?
3. Assume there are 50,000 Ampay dealers in the United States, and each is supposed to recruit six more dealers. These in turn are supposed to recruit six more dealers each, and so on. (a) How many completely successful levels of recruitment would assure that the average family in the United States would have at least one Ampay dealer? (b) How many levels of recruitment would assure that every man, woman and child in the United States would be an Ampay dealer? (c) Now, allowing overseas dealerships, how many levels of recruitment would result in every human being on earth becoming an Ampay dealer?
4. The radioactive isotope strontium 90 has a half-life of 28.1 years. How much time must elapse after an atomic explosion before the resulting strontium 90 decays to one-eighth of its initial amount?
5. A tank storing solar-heated water stands unmolested* in a room having an approximately constant temperature of 80°F. If the tank cools from 120°F to 100°F

This solar home uses two 4000-gallon water tanks for heat storage.

in 3 days, (a) what will be its temperature after 3 more days? (b) When will it reach 85°F? (c) When will it reach 80°F?
[*"Unmolested" means no heat is added, and no water is added or removed during the period under consideration.]

6. A cup of coffee in a room where the temperature is 70°F cools from 190°F to 130°F in 6 minutes. How much longer will it take to cool down to 85°F?

7. Assume you have $100 for gambling at the roulette wheel, and at each spin of the wheel you bet half the money you have then on "red," which is an "even-money" bet.
   (a) Show that whenever you win, your funds are multiplied by $\frac{3}{2}$, and whenever you lose, your funds are multiplied by $\frac{1}{2}$.
   (b) Show that n wins and n losses (in any order) multiply your funds by $(\frac{3}{4})^n$. [*Hint*. Start with $n = 1$; and then $n = 2$.]
   (c) Show that if you win 10 times and lose 10 times you will have only about $5.63 left.
   (d) Does the operator of the roulette wheel (say the casino) really get your money this easily? (If so, why do casinos bother to adjust their games to assure a profit for the casino?)

8. If you like your coffee hot, should you add the cream at once when the coffee is poured, or should you wait until just before you will drink it?

6.2. More on Half-Lives

The examples and problems in Section 6.1 dealing with radioactive decay and Newton's law of cooling were formulated with numbers that "come out nicely." Specifically, the only times involved were integer multiples of the half-life.

In each case there is a certain decaying quantity, say Q, to be considered. This might be the amount of some radioactive material, or the difference between the temperature of a heated object and the ambient temperature. Assume one starts with an initial value $Q = Q_0$. Then after a certain time h, Q is reduced to $\frac{1}{2}Q_0$. This h is the half-life. After an *additional* time h one finds $Q = \frac{1}{4}Q_0$, and so on. Thus the values of Q at the ends of successive intervals of time of length h form the geometric sequence

$$Q_0, \quad Q_0(\tfrac{1}{2}), \quad Q_0(\tfrac{1}{2})^2, \quad Q_0(\tfrac{1}{2})^3, \ldots .$$

Another way of expressing this is to say that the value of Q after time t is

$$Q = Q_0(\tfrac{1}{2})^{t/h} \qquad (*)$$

provided $t = 0$, or h, or $2h$, or $3h$, ... Take time to confirm this by substituting into (*) $t = 0$, $t = h$, and $t = 2h$.

But what can be said if $t = \frac{1}{2}h$ or $1.2h$ or some other noninteger multiple of h? Consider the following specifically.

Example 1. Radium 226 has a half-life $h = 1620$ years. If one starts with 10 grams of radium 226, how much will remain after 810 years? (This is a continuation of Example 2 of Section 6.1.)

Solution. Let Q = number of grams of radium 226 remaining after t years.
Then for $t = 0, 1620, 3240, \ldots$

$$Q = 10(\tfrac{1}{2})^{t/1620}.$$

Even though it has not been justified, let us try to use this relation when $t = 810$ years. This would give

$$Q = 10(\tfrac{1}{2})^{t/1620} = 10(\tfrac{1}{2})^{810/1620} = 10(\tfrac{1}{2})^{1/2}.$$

But what is the meaning of $(\frac{1}{2})^{1/2}$?

More generally, what would be a reasonable interpretation of the symbol $r^{1/2}$? The clue for defining this "fractional exponent" is found in the laws of exponents. *If* the usual laws of exponents are still to hold after one defines $r^{1/2}$, then one must have

$$(r^{1/2})^2 = r^{(1/2)\cdot 2} = r.$$

Thus it seems natural to define

$$r^{1/2} = \sqrt{r}.$$

Returning to the question which gave rise to this discussion, one is led to conjecture that after 810 years there will remain

$$Q = 10\sqrt{\frac{1}{2}} = 10\left(\frac{1}{\sqrt{2}}\right) \text{ grams of radium 226.}$$

But is this correct? It says that after 810 years the amount is reduced to $1/\sqrt{2}$ times its initial value. If this is true, then in another 810 years it should be reduced to $(1/\sqrt{2})(1/\sqrt{2}) = \frac{1}{2}$ of the original amount. And this is exactly as it should be after a total of 1620 years! So we conclude with confidence that after 810 years there will remain

$$\frac{10}{\sqrt{2}} = \frac{10}{\sqrt{2}} \cdot \frac{\sqrt{2}}{\sqrt{2}} = \frac{10\sqrt{2}}{2} = 7.07 \text{ grams.}$$

In an introductory course in calculus, one learns how to define r^p for any $r > 0$ and any (real) number p. And indeed, the "correct" definition for $r^{1/2}$ does turn out to be $\sqrt{r}$. For reference, Figure 1 gives a graph of the values of $(\frac{1}{2})^p$ for values of p from 0 to 6.5. Such a calculus course would also include a more complete discussion of radioactive decay showing the validity of (*) for arbitrary values of $t \geq 0$, not just integer multiples of h.

An important application of this discussion is the science of "carbon dating" due to Willard Libby around 1949. Any sample of carbon is composed *in part* of radioactive carbon 14, which has a half-life of about 5700 years. But while the carbon 14 is decaying (into nitrogen), more is continually

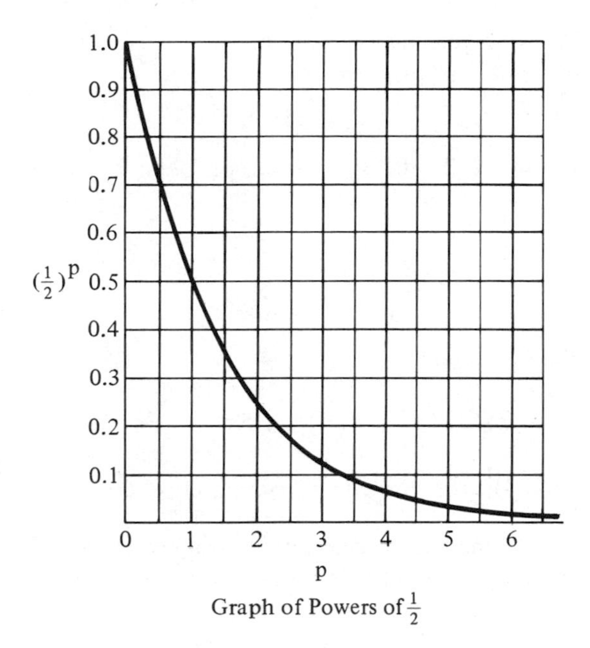

Graph of Powers of $\frac{1}{2}$

Figure 1.

being produced by the bombardment of the earth's atmosphere by cosmic rays. This results in a constant equilibrium level for the *percentage* of carbon 14 in a carbon sample from the earth's atmosphere. Indeed, this equilibrium level is assumed to have been constant for thousands of years.

Plants get their carbon from the atmosphere, and so do animals—perhaps by eating the plants or other animals. So as long as a plant or animal is *living* and continuing to exchange carbon with the atmosphere, the ratio of carbon 14 to ordinary carbon in that living organism should be that of the constant equilibrium level of the atmosphere. But when a plant or animal dies it no longer receives carbon from the environment, and its carbon 14 decays without replacement. This provides a basis for determining how long ago that plant or animal died.

Example 2. If the wooden beams from a certain cliff dwelling contain only 25% of the carbon 14 found in a comparable sample of living wood, how long ago were these beams cut?

Solution. The carbon 14 in these beams decayed to half its original amount in $h = 5700$ years. So if the carbon 14 is found to be just one-fourth of its original amount, the beams must have been cut approximately $2h = 11{,}400$ years ago.

Example 3. If the carbon (charcoal) from an ancient campfire contains only 35% of the carbon 14 found in a comparable sample of living wood, how long ago was this wood burned?

Solution. If Q_0 was the original amount of carbon 14 in the sample, then by (*), the amount remaining after t years is

$$Q = Q_0(\tfrac{1}{2})^{t/5700}.$$

In the present example $Q = 0.35Q_0$. So the problem is to find t such that

$$(\tfrac{1}{2})^{t/5700} = 0.35.$$

Reference to Figure 1 shows that $(\frac{1}{2})^{1.5} = 0.35$ approximately. Thus

$$(\tfrac{1}{2})^{t/5700} \cong (\tfrac{1}{2})^{1.5}.$$

So, equating the two exponents,

$$\frac{t}{5700} \cong 1.5$$

or $t \cong 8550$ years. The fire burned approximately 8550 years ago.

Example 4. A cup of instant coffee is mixed using water at 195°F, and then is left to stand in a room where the temperature is 75°F. If the coffee cools to 105°F in 20 minutes, when will it reach 95°F?

Solution. Let T = the temperature of the coffee in degrees Fahrenheit at time t minutes after mixing.

Then (*) becomes $T - T_a = (T_0 - T_a)(\frac{1}{2})^{t/h}$, or

$$T - 75 = (195 - 75)(\tfrac{1}{2})^{t/h},$$

where h is the half-life of the temperature difference in minutes. To determine the value of h, use the information that $T = 105$ when $t = 20$. Thus

$$105 - 75 = (195 - 75)(\tfrac{1}{2})^{20/h}$$

$$\therefore 30 = 120(\tfrac{1}{2})^{20/h}$$

$$\therefore \left(\frac{1}{2}\right)^{20/h} = \frac{30}{120} = \frac{1}{4} = \left(\frac{1}{2}\right)^2.$$

Now, in order that $(\frac{1}{2})^{20/h} = (\frac{1}{2})^2$, it is necessary that

$$\frac{20}{h} = 2.$$

So $h = 10$ minutes, and the relation between T and t can be rewritten as

$$T - 75 = (195 - 75)(\tfrac{1}{2})^{t/10}.$$

To answer the question, "when will $T = 95$?," one substitutes $T = 95$ into the above to find

$$95 - 75 = (195 - 75)(\tfrac{1}{2})^{t/10}.$$

From this,

$$20 = 120(\tfrac{1}{2})^{t/10}$$

or

$$(\tfrac{1}{2})^{t/10} = \tfrac{1}{6} = 0.1666\ldots.$$

Here it is necessary to refer to Figure 1 to estimate the power p for which $(\frac{1}{2})^p = 0.1666\ldots.$ It appears that $p \cong 2.6$. Thus

$$\frac{t}{10} \cong 2.6,$$

and $t \cong 26$ minutes after the coffee was mixed.

Example 5. A saucepan of soup is removed from the stove at 212°F and left to stand in a room where the temperature is 68°F. If the soup cools to 111°F in 35 minutes, when was it 140°F?

Solution. Let T be the temperature of the pan of soup in degrees Fahrenheit t minutes after removal from the stove.

Then

$$T - 68 = (212 - 68)(\tfrac{1}{2})^{t/h}$$

where h is the half-life in minutes. Substitute $T = 111$ and $t = 35$ to find

$$\begin{aligned} 111 - 68 &= (212 - 68)(\tfrac{1}{2})^{35/h} \\ \therefore 43 &= 144(\tfrac{1}{2})^{35/h} \\ \therefore \left(\frac{1}{2}\right)^{35/h} &= \frac{43}{144} \cong 0.3. \end{aligned}$$

And Figure 1 then shows that $35/h \cong 1.75$. Hence $h \cong 20$ minutes.

Now, to find the value of t at which $T = 140$, one must solve

$$140 - 68 = (212 - 68)(\tfrac{1}{2})^{t/20}$$

or

$$72 = 144(\tfrac{1}{2})^{t/20}.$$

This is equivalent to

$$(\tfrac{1}{2})^{t/20} = \tfrac{1}{2}.$$

So $t/20 = 1$, or $t = 20$ minutes (approximately).

Problems

1. The discussion of Example 1 led to the definition $(\frac{1}{2})^{0.5} = 1/\sqrt{2} = 0.707$. (a) If the laws of exponents are to be preserved, how should one define $r^{1.5}$ and $r^{2.5}$? (b) Specifically, find $(\frac{1}{2})^{1.5}$ and $(\frac{1}{2})^{2.5}$, and compare your answers with the corresponding points on Figure 1.
2. (a) How should one define $r^{0.25}$? (b) Compute $(\frac{1}{2})^{0.25}$ and check your answer with Figure 1.

3. If samples of carbon from the cloth wrapping an Egyptian mummy contain only 60% as much carbon 14 as in a comparable sample of carbon from a living plant, when did this mummified individual die? (Recall, carbon 14 has a half-life of 5700 years.)

4. Plutonium 239, one of the products of atomic explosions and (fission) nuclear power plants, has a half-life of about 24,400 years. If a nuclear war or an accident at a nuclear power plant deposited 10 times the "acceptable" level of plutonium 239 on some locality, how long would it take to decay to the acceptable level?

5. A tank storing solar-heated water stands unmolested in a room having an approximately constant temperature of 80°F. (Compare with Problem 5 of Section 6.1.) If the tank cools from 120°F to 100°F in 3 days, what will be its temperature after one more day?

6. Assume that you have a power failure. Your house is at 65°F, but your furnace will no longer run. The temperature outside is 49°F, and in 4 hours your house cools down to 61°F. What will be the indoor temperature after 4 more hours? (Assume the outdoor temperature remains at 49°F and ignore the fact that the ground temperature may be somewhat different.)

7. In the 1940s radioactive residues from the Manhattan Project, which developed the atomic bomb, were deposited on a 1500-acre site at Lewiston, N.Y. (near Lake Ontario, 10 miles north of Niagara Falls). In 1972 large amounts of earth were removed to "decontaminate" parts of the site. But in 1980 samples of material from the central ditch still showed 332 times the concentration of radium 226 considered safe by the State of New York. (*New York Times*, June 23, 1980, page B4.) Recalling that the half-life of radium 226 is 1620 years, how long would it take for this to decay to a safe level assuming no more were removed? [*Hint*. At some point you will need to use both Figure 1 *and* a law of exponents to find p such that $(\frac{1}{2})^p = 1/332$.]

8. Some people say that you can save fuel by turning down the thermostat at night. Others argue that this is no use since it will take so much more fuel to reheat the house in the morning. Which do you believe? Why?

9. During the hot summer months of 1980 several supermarkets posted signs saying they were keeping their stores cooler than the federal-guideline 78°F because otherwise their food freezers would have to work harder and this would waste even more energy. What do you think? Why?

6.3. Compound Interest and Related Matters

If an amount of money, say P dollars, is deposited in a savings account and left there, interest accumulates on P (the principal). And after the interest is "credited to the account," further interest will accumulate on both P and the previous interest.

Assume the account pays $I\%$ per year compounded annually. Then the first year's interest on the initial deposit P will be

$$P \times \frac{I}{100} \quad \text{dollars.}$$

To express this more compactly, define $i = I/100$. Then the first year's interest can be written as

$$Pi \quad \text{dollars.}$$

Thus at the end of the first year the total amount in the account is the initial deposit (the principal) plus the interest,

$$P + Pi = P(1 + i).$$

Now, assuming that nothing else is deposited in or withdrawn from the account, the interest during the second year is computed on the new principal $P(1 + i)$. Hence, during the second year the interest earned is $P(1 + i)i$. So the total balance in the account at the end of the second year is

$$P(1 + i) + P(1 + i)i.$$

This is written more compactly as

$$P(1 + i)(1 + i) = P(1 + i)^2.$$

During the third year the interest, computed on this latter balance, is $P(1 + i)^2 i$. So, after 3 full years, the balance will be

$$P(1 + i)^2 + P(1 + i)^2 i = P(1 + i)^3.$$

Similarly, after 4 years it will be

$$P(1 + i)^4,$$

and after n years the balance becomes

$$P(1+i)^n.$$

This looks like ar^n again, as in Section 6.1. Here $a = P$ and $r = 1 + i$.

Example 1. Assume that \$15.00 is deposited and left in a savings account which pays 4% interest per year compounded annually. What will be the balance in the account after 5 years?

Solution. Since $i = 4/100 = 0.04$, the answer is

$$15(1+0.04)^5 = 15(1.04)^5 = 18.25 \text{ dollars.}$$

This value has been rounded off to the nearest cent.

It is more customary in savings accounts for the interest to be compounded quarterly rather than annually. In this case one finds that the balance in an account after n *quarters* is

$$P(1+i)^n,$$

where i is now obtained from the stated annual interest rate by *dividing by 4*—the number of quarters in a year.

Example 2. Assume that \$15.00 is deposited and left in a savings account which pays 4% interest per year compounded quarterly. What will be the balance after 5 years?

Solution. Since $i = 1/100 = 0.01$, the amount in the account after $n = 20$ quarters (5 years) is

$$15(1+0.01)^{20} = 15(1.01)^{20}.$$

But how can one calculate $(1.01)^{20}$ without a great deal of work?

Even with the simplest hand calculator, you can find r^{20} fairly efficiently as follows. Note that $r^{20} = (r^4)^5 = (r^4)^4(r^4)$. Now almost any calculator will square numbers easily. So $r^4 = (r^2)^2$ is found quickly. Put the result in the calculator's "memory." Then compute $(r^4)^4$ by two more squarings, and finally multiply by r^4 from the memory.

Do this with $r = 1.01$; and finally multiply by 15. You should find

$$15(1.01)^{20} = 18.30 \text{ dollars}$$

(rounded off to the nearest cent).

Note that the more frequent compounding results in slightly more interest than in Example 1.

Computational tricks similar to that used above for $(1.01)^{20}$ will be needed again. (*Exception.* If your calculator has a "power" key, it can produce r^n more quickly. But you may not learn as much doing it that way.)

The same type of calculation used for compound interest is involved in estimates of growth for inflation and for populations.

Example 3. In mid 1983 the world population was estimated to be 4.72 billion and increasing at the rate of 1.8% per year. Forecast the world population for mid 1990.

Solution. This is just like a "deposit" of 4.72 billion earning "interest" at 1.8% per year compounded annually. So, as in the discussion and examples above, the population after 7 years would be

$$4.72(1 + 0.018)^7 = 4.72(1.018)^7 = 4.72(1.133) = 5.35 \text{ billion.}$$

The next example is a compound interest problem to be solved "backwards."

Example 4. If the value of an investment increases by 26% in 2 years, what is the equivalent effective annual rate of interest? (No, it is not 13%.)

Solution. Let $I\%$ be the equivalent effective annual rate of interest—assumed to be compounded annually. Let $i = I/100$. Then in 2 years an original deposit or investment of P dollars grows to

$$P(1 + i)^2 \text{ dollars.}$$

But this is to be P increased by 26%, namely, $P(1.26)$. Thus

$$(1 + i)^2 = 1.26.$$

So

$$1 + i = \sqrt{1.26}.$$

Using the method of Section 3.3, you can find $\sqrt{1.26} = 1.1225$. Therefore, $i = 0.1225$, for an effective annual interest rate of 12.25% per year.

A zero coupon CD (certificate of deposit) is a savings certificate which offers no current interest but is sold at a discount from its value at "maturity" on some future date. So the "interest" is represented by the difference between the cost of the certificate and its value at maturity. In simple cases, if you know the cost of the certificate, its maturity value, and the time until maturity, you can now determine the effective interest rate.

Example 5. Assume a zero coupon CD costs \$849.46 today and will mature at \$1000 in 2 years. Find the effective annual rate of interest, assuming annual compounding.

Solution. Let I be the effective annual rate of interest in percent, and let $i = I/100$. Then a deposit of \$849.46 accumulating interest for 2 years gives a balance of

$$\$849.46(1 + i)^2.$$

This must be the maturity value. Thus

$$849.46(1 + i)^2 = 1000.$$

So $(1 + i)^2 = 1000/849.46 = 1.1772$. Taking square roots, one finds

$$1 + i = \sqrt{1.1772} = 1.085.$$

(This square root can be calculated by the method of Section 3.3.) Thus $i = 0.085$, for an effective annual interest rate of 8.5%.

Example 6. Find the effective annual rate of interest, assuming annual compounding, on a zero coupon CD which costs $500 and matures in 4 years to $700.62.

Solution. Again, let $I\%$ be the effective annual rate of interest, and let $i = I/100$. Then

$$500(1 + i)^4 = 700.62.$$

So $(1 + i)^4 = 700.62/500 = 1.4012$. Taking square roots, one finds

$$(1 + i)^2 = \sqrt{1.4012} = 1.1837.$$

Take square roots once more to find

$$1 + i = \sqrt{1.1837} = 1.088.$$

Thus the effective annual interest rate is 8.8%.

If you wanted to find the effective annual interest rate for a zero coupon CD which matures in 3 years, you would need to compute a cube root. This will become possible in Section 7.4.

Problems

1. If 5 dollars earns 6% interest per year, what will be the accumulated total after 10 years if the interest is (a) not compounded? (b) compounded annually? (c) compounded quarterly? [*Hint.* $r^{40} = (r^8)^4 r^8$.]

2. One thousand dollars is deposited and left in an account which pays interest at an annual rate of 10% compounded semiannually. Determine the balance after 6 years.

3. If $1000 earns interest at the rate of 10% per year, what will be the accumulated total after 20 years if the interest is (a) not compounded? (b) compounded annually? (c) compounded quarterly?

4. If the "Consumer Price Index" increases 1% each month for a year, by how much does it increase for the entire year?

5. The population of India in 1980 was almost 700 million. If the population continued to increase at the rate of 2% per year, what would it be in the year 2000?

6. If the population of a certain country grows at a constant rate of 2% per year, show that it will double in 35 years. [*Hint*. In the actual computation you might use the fact that $r^{35} = r^{32} \cdot r \cdot r \cdot r$, and r^{32} is easy to compute with even a simple calculator.]

7. Assuming the validity of the data in Problem 5, and assuming this same annual growth rate continues, (a) when would the population of India reach four billion? (b) What would be the population in the year 2105?

8. A zero coupon CD costing $1000 today matures to $1250 in 2 years. Find the effective annual rate of interest, assuming annual compounding.

9. A zero coupon CD costing $703.25 today will mature to $1000 in 4 years. Find the effective annual rate of interest if compounded annually.

* 10. Investment Company A imposes a 6% sales charge when you buy shares in its investment program. Investment Company B has no initial sales charge but it deducts a withdrawal charge of 6% when you cash in your shares. Each company pays 10% interest per year compounded quarterly on the moneys invested with it. And the sales representatives each insist that their company's system of charges is better. Show that the result will be the same either way.

11. Banks have different ways of compounding interest. Assume that a bank advertises a nominal annual interest rate of 10%. Find the effective annual rate (the annual yield) if the bank (a) compounds quarterly, (b) compounds monthly, (c) compounds daily, (d) computes the daily rate as though there were only 360 days in a year and then compounds the interest for 365 days. [Some banks do actually use the latter method.]

* 12. Determine the method of compounding used by banks which advertise
    (a) a nominal rate of 10.2% and an annual yield of 10.69%,
    (b) a nominal rate of 10.1% and an annual yield of 10.78%,
    (c) a nominal rate of 10.2% and an annual yield of 10.60%.

6.4. IRAs and Similar Tax Sheltered Accounts

Various "tax sheltered" retirement plans are available to certain classes of Americans. Self-employed persons can have "Keogh plans," while employees of public schools and certain tax-exempt organizations are eligible for "Tax Sheltered Annuities," or "Supplemental Retirement Accounts."

Beginning in 1982 the IRA (Individual Retirement Account) became available to *all* persons with "earned income."

This section concentrates on the possible income tax benefits of an IRA—the most universal plan. The other plans mentioned above work in similar ways.

An IRA is an investment account permitting an individual to defer income taxes on both the earnings deposited and the interest (or dividends) left to accumulate in the account. The investor must pay taxes on both the principal and interest when the proceeds of the account are withdrawn after retirement. IRAs are offered by banks, stock brokers, and insurance companies.

The institutions and salespersons offering such investment plans often say that an IRA will be to your advantage (as compared to an ordinary savings account) because:

(1) you will earn interest on money which would have gone as taxes, and
(2) when you do pay the taxes at retirement, you will probably be in a lower income tax bracket because of lower income.

Reason (2) cannot be analyzed mathematically. Who knows how the tax laws might change between now and the time of your retirement? And even if the laws remain unchanged, there is no assurance that your income will be significantly reduced upon retirement.

So let us suppose that your tax bracket remains unchanged, and examine reason (1). Recalling Example 8 of Section 1.3 and Problem 10 of Section 6.3, you might reasonably suppose that it makes no difference whether you pay tax on the initial deposit and the accumulating interest as you go, or wait until retirement and pay taxes on the total proceeds. But now there *is* a difference as the following example indicates.

Example 1. A wage earner in the 40% tax bracket wants to commit \$2000 of her *before-tax earnings* to an investment account. Compare the results of the following two alternatives after 10 years. (a) She could put the entire \$2000 into an IRA paying 10% per year compounded annually, and let the interest accumulate. In this case she would defer taxes on the principal and interest until the money is withdrawn after retirement. Or (b) she could pay the 40% tax, and invest the remaining \$1200 in a conventional account paying 10% per year compounded annually. In that case, she would have to pay taxes on the interest earned each year even if none were withdrawn from the account.

Solution. Assume that the rate of interest remains at 10% for both accounts and that the investor's tax bracket remains at 40%. (In reality, these numbers will vary and cannot be predicted. But one must assume *something* in order to do any calculations.)

(a) The IRA starts with the full \$2000 and interest accumulates at 10% per year—with no taxes paid until withdrawal. Thus, after 10 years the account shows a balance of

$$\$2000(1 + 0.1)^{10} = \$2000(1.1)^{10} = \$5187.48.$$

(The value of $(1.1)^{10}$ can be obtained using a simple calculator as in Section 6.3. Check it!)

Now the funds are withdrawn and the 40% tax is paid, leaving

$$\$5187.48(1 - 0.4) = \$3112.49.$$

(b) The conventional "taxed-as-you-go" account starts with \$1200 (\$2000 less the 40% tax). In this case the investor will owe 40% of the interest earned each year as taxes; so a fair comparison with the IRA requires that she take this money out of the account each year to pay the tax collector. Thus the effective annual interest rate is reduced to 6% (60% of 10%). After 10 years the account will yield

$$\$1200(1 + 0.06)^{10} = \$1200(1.06)^{10} = \$2149.02$$

with all taxes paid. (Check this with a calculator.)

It is now clear that the investor would be much better off with the IRA.

Look back at the IRA calculation in part (a) above. Notice that the final return after paying the tax is

$$\$2000(1 + 0.1)^{10}(1 - 0.4).$$

By the commutative law of multiplication this is equivalent to

$$\$2000(1 - 0.4)(1 + 0.1)^{10} = \$1200(1 + 0.1)^{10}.$$

In other words, the IRA effectively took the after-tax earnings of \$1200 and paid interest at 10% per year compounded annually and tax free.

One example does not prove that an IRA will *always* be the better choice. The following theorem generalizes the example.

Theorem. *If an investor's tax rate remains fixed, an IRA effectively makes the accumulated interest tax free as compared to an ordinary investment account. Tax is paid on the original earnings in both cases.* (*It is assumed that the funds are left in the account until the investor retires.*)

PROOF. For simplicity, the proof will be given for the special case of a fixed interest rate. In Problem 9 the interested reader is invited to improve the proof, removing this irrelevant hypothesis.

Assume that the investor has a tax rate of $T\%$ which remains constant; and assume that the interest rate is $I\%$, also remaining constant. *Before-tax earnings* of \$2000 are committed to the investment account for n years, at which time (after retirement) the investor withdraws the proceeds. Let $t = T/100$ and $i = I/100$

(a) In the IRA, the full \$2000 is invested, and the interest rate is the full $I\%$. So, after n years, the account shows a balance of

$$\$2000(1 + i)^n.$$

Upon withdrawal the investor pays taxes, leaving net proceeds of

$$\$2000(1 + i)^n(1 - t) \qquad \text{with all taxes paid.}$$

(b) In the ordinary taxed-as-you-go account, the net investment is $\$2000(1 - t)$—the balance after paying taxes on the \$2000 earnings. Then, since taxes must be paid on the interest each year, the effective annual interest rate is reduced to $I(1 - t)\%$. So the proceeds of the investment after n years will be

$$\$2000(1 - t)[1 + i(1 - t)]^n \quad \text{with all taxes paid.}$$

To compare the two results, apply the commutative law of multiplication to the result of the IRA calculation, expressing the net proceeds as

$$\$2000(1 - t)(1 + i)^n.$$

Then you can see that both cases (a) and (b) can be regarded as accounts paying interest on an initial deposit of $\$2000(1 - t)$. In the ordinary account of case (b), the effective interest rate is reduced by taxes to $I(1 - t)\%$. But the IRA considered in (a) effectively pays interest at an annual rate of $I\%$—i.e. tax free. □

In both the example and the theorem, the amount deposited in the IRA was taken as \$2000. That is because \$2000 was the maximum annual deposit allowed in an IRA for a single wage-earner in the tax laws which took effect in 1982. The results for a larger or smaller deposit in a tax sheltered investment account can be found in exactly the same way.

If the investor finds himself or herself in a lower tax bracket after retirement, the benefit of deferring taxes would be further enhanced. However, as the theorem shows, even without assuming any change in tax bracket the IRA offers a big advantage.

Many institutions and salespersons offering IRAs seem to be unaware of this theorem. If they knew it, they could make a much stronger case for IRAs while omitting the prattle about "probable lower income tax brackets at retirement."

But suppose you are a long way from retirement and decide to withdraw the funds from your IRA. Any funds withdrawn before age $59\frac{1}{2}$ are subject to a 10% *penalty* in addition to ordinary income tax. So, if you are going to withdraw funds before you reach $59\frac{1}{2}$, it is no longer obvious that an IRA is the better choice.

Example 2. The wage earner of Example 1 leaves her \$2000 in an IRA earning 10% per year compounded annually for 10 years. But when she withdraws the proceeds she is not yet $59\frac{1}{2}$ years old. How much does she get after taxes in this case?

Solution. As in the solution of Example 1, the account shows a balance of \$5187.48 at the time of withdrawal. But now she must pay a 10% penalty in addition to her 40% tax. This leaves

$$\$5187.48(1 - 0.5) = \$2593.74$$

after taxes. This is still an improvement on the return from the comparable taxed-as-you-go account—but not as dramatic as in Example 1.

Example 3. A wage earner in the 25% tax bracket wants to use \$1000 of before-tax earnings for an investment account. Compare the results of the following two alternatives if the money is withdrawn after only 3 years when the investor is still under $59\frac{1}{2}$. (a) He puts the entire amount in an IRA paying 9% per year compounded annually and lets the interest accumulate. Or (b) he pays the tax on the \$1000 and puts the remaining \$750 in a conventional savings account paying 9% per year compounded annually.

Solution. (a) Using the IRA, the account will show a balance after 3 years of

$$1000(1 + 0.09)^3 = 1295.03 \text{ dollars.}$$

Upon withdrawal the investor must pay the 10% penalty plus his regular 25% tax—reducing the proceeds by 35%. The result is

$$\$1295.03(1 - 0.35) = \$841.77.$$

(b) In the other option he invests only \$750 and his interest is effectively reduced by taxes to 75% of 9%, namely, 6.75%. So after 3 years the proceeds, tax paid, will be

$$\$750(1 + 0.0675)^3 = \$912.36.$$

In this case the investor is better off with the ordinary account.

Problems

[*Note.* In all questions involving ordinary (taxed-as-you-go) accounts, you should imagine that enough money is withdrawn from the account each year to pay the required tax on the interest (as was done in the examples).]

1. A 28-year-old wage earner in the 25% tax bracket commits $2000 of before-tax earnings to an investment account paying 8% interest compounded annually. Assume that these rates remain constant. Find the after-tax proceeds if the money were left in the account for 35 years (a) in an IRA, (b) in an ordinary account.

2. Repeat Problem 1 for a wage earner in the 40% tax bracket and investment accounts paying 10% interest compounded annually.

3. A wage earner in the 30% tax bracket commits $1000 of before-tax earnings to an investment account paying 9% interest compounded annually. He withdraws the funds after retirement 12 years later at age 70. At this time he has dropped to the 25% tax bracket (due to suddenly decreased income). Compare the after-tax proceeds if (a) the funds were in an IRA and (b) the funds were in an ordinary savings account.

4. Repeat Problem 3 assuming that the retiree now suddenly finds himself in the 35% tax bracket (due to the success of a small business he had started).

5. At age 27 you deposit $500 in an IRA paying 9% per year compounded annually. You withdraw this money when you urgently need cash 5 years later. Assuming that your tax bracket remains at 25% throughout this time, (a) what are the proceeds? (b) What would they have been from a comparable investment in an ordinary account paying 9%?

6. Repeat Problem 5 assuming that you leave the funds in the account(s) for 8 years.

7. Repeat Problem 5 assuming an 8-year period and a 40% tax bracket.

*8. You are opening an IRA with the full expectation that you will be withdrawing the funds before age $59\frac{1}{2}$ and hence will have to pay the 10% penalty (as in Problems 5, 6, and 7). Show that in this case (compared to an ordinary savings account) the IRA becomes more attractive the higher the interest rate, the higher your tax bracket, and the longer the funds stay in the account.

*9. Prove the theorem on page 93 *without* assuming a fixed interest rate.

6.5. Geometric Series—The "Sum" of a Geometric Sequence

Section 6.1 introduced the concept of a geometric sequence

$$a, ar, ar^2, ar^3, \ldots.$$

Now the question will be, can one make any sense of the "sum" of all the terms of such a sequence? In other words, can one give a sensible definition for the expression, referred to as a **geometric series,**

$$a + ar + ar^2 + ar^3 + \cdots ?$$

The trouble is that *addition* has only been defined for a *finite* number of terms.

Example 1. Try to assign a reasonable meaning to the sum of the series

$$1 + \tfrac{1}{3} + (\tfrac{1}{3})^2 + (\tfrac{1}{3})^3 + \cdots.$$

This is a geometric series with $a = 1$ and $r = \frac{1}{3}$.

Solution. The idea will be to first examine the sum of a large but *finite* number of terms of the series. Specifically, consider the sum of the terms from 1 through $(\frac{1}{3})^n$, calling this sum S_n. Thus

$$S_n = 1 + \tfrac{1}{3} + (\tfrac{1}{3})^2 + (\tfrac{1}{3})^3 + \cdots + (\tfrac{1}{3})^n.$$

There is a simple but clever trick for computing this sum. Write out the expression for $\frac{1}{3}S_n$ (r times S_n) directly below S_n itself, aligning like powers of $\frac{1}{3}$. Then subtract. This gives

$$\begin{aligned} S_n &= 1 + \tfrac{1}{3} + (\tfrac{1}{3})^2 + (\tfrac{1}{3})^3 + \cdots + (\tfrac{1}{3})^n \\ \tfrac{1}{3}S_n &= \quad\;\; \tfrac{1}{3} + (\tfrac{1}{3})^2 + (\tfrac{1}{3})^3 + \cdots + (\tfrac{1}{3})^n + (\tfrac{1}{3})^{n+1} \\ \hline \tfrac{2}{3}S_n &= 1 \qquad\qquad\qquad\qquad\qquad\qquad\qquad - (\tfrac{1}{3})^{n+1}. \end{aligned}$$

It remains only to divide through by $\frac{2}{3}$ to find

$$S_n = \tfrac{3}{2} - \tfrac{3}{2}(\tfrac{1}{3})^{n+1}.$$

Does this expression for the sum of a finite number of terms ($n + 1$ terms to be exact) now suggest how one ought to define the sum of the infinitely many terms of the entire original geometric series?

Consider what happens to S_n as n increases, i.e., as one includes more and more terms in the finite sum represented by S_n. Successively higher powers of $\frac{1}{3}$ yield successively smaller numbers. Thus

$$\tfrac{1}{3} = \tfrac{1}{3}, \qquad (\tfrac{1}{3})^2 = \tfrac{1}{9}, \qquad (\tfrac{1}{3})^3 = \tfrac{1}{27}, \quad (\tfrac{1}{3})^4 = \tfrac{1}{81}, \cdots.$$

By choosing n sufficiently large, you can make $(\frac{1}{3})^n$ or $(\frac{1}{3})^{n+1}$ as small as you please. So $(\frac{1}{3})^{n+1}$ is said to "approach zero" as n "approaches infinity." Hence

$$S_n = \tfrac{3}{2} - \tfrac{3}{2}(\tfrac{1}{3})^{n+1} \quad \text{"approaches } \tfrac{3}{2}\text{"}$$

as n "approaches infinity." This value, $\frac{3}{2}$, will be called the *sum* of the original geometric series; and one writes

$$1 + \tfrac{1}{3} + (\tfrac{1}{3})^2 + (\tfrac{1}{3})^3 + \cdots = \tfrac{3}{2}.$$

The solution of Example 1 serves as a model for defining the sum of a general geometric series,

$$a + ar + ar^2 + ar^3 + \cdots.$$

First examine the sum of finitely many terms

$$S_n = a + ar + ar^2 + \cdots + ar^n.$$

As in Example 1, subtract rS_n from S_n. Thus

$$\begin{array}{rl} S_n = & a + ar + ar^2 + ar^3 + \cdots + ar^n \\ rS_n = & \quad\; ar + ar^2 + ar^3 + \cdots + ar^n + ar^{n+1} \\ \hline S_n - rS_n = & a \qquad\qquad\qquad\qquad\qquad\qquad\quad - ar^{n+1} \end{array}$$

or

$$(1 - r)S_n = a - ar^{n+1}.$$

Division by $1 - r$ is permissible if and only if $r \neq 1$. Thus

$$S_n = \frac{a}{1-r} - \frac{a}{1-r} r^{n+1} \qquad \text{provided } r \neq 1.$$

In many instances one has $0 < r < 1$. It is then possible to give a reasonable interpretation to the "infinite sum"

$$a + ar + ar^2 + ar^3 + \cdots.$$

Start with the finite sum $S_n = a + ar + ar^2 + \cdots + ar^n$ and ask what happens as one adds more and more terms, i.e., as n approaches infinity. Since $0 < r < 1$, the powers of r, namely, r^2, r^3, r^4, ..., become progressively smaller approaching zero. (Why?) Thus r^n approaches zero as n approaches infinity. The expression under consideration is

$$S_n = \frac{a}{1-r} - \frac{a}{1-r} r^{n+1}$$

and, since r^{n+1} approaches zero, S_n approaches

$$S = \frac{a}{1-r}.$$

This is then called the **sum** of the geometric series

$$a + ar + ar^2 + ar^3 + \cdots.$$

Example 2. Interpret the repeating decimal

$$0.\overline{3} = 0.3333\ldots$$

as a geometric series, and find the sum of that series.

Solution. The number 0.3333... represents

$$\frac{3}{10} + \frac{3}{100} + \frac{3}{1000} + \frac{3}{10{,}000} + \cdots$$
$$= 3\left(\frac{1}{10}\right) + 3\left(\frac{1}{10}\right)^2 + 3\left(\frac{1}{10}\right)^3 + 3\left(\frac{1}{10}\right)^4 + \cdots.$$

This is a geometric series with $a = \frac{3}{10}$ and $r = \frac{1}{10}$. So the sum is

$$S = \frac{a}{1-r} = \frac{3/10}{1 - 1/10} = \frac{3}{10-1} = \frac{1}{3}.$$

This is the familiar equivalent fraction for 0.3333. . . .

This is an alternative approach to the problem of converting a repeating decimal to a fraction. (Compare with Section 2.3.)

Example 3. The repeating decimal

$$4.2\overline{13} = 4.2131313\ldots$$

is 4.2 plus the geometric series

$$0.0131313\ldots = 0.013 + 0.013(\tfrac{1}{100}) + 0.013(\tfrac{1}{100})^2 + \cdots.$$

Here $a = 0.013$ and $r = \frac{1}{100}$. Thus

$$4.2\overline{13} = 4.2 + \frac{a}{1-r} = 4.2 + \frac{0.013}{1-0.01} = 4 + \frac{2}{10} + \frac{13}{990} = 4\frac{211}{990}.$$

Returning to the general case

$$a + ar + ar^2 + ar^3 + \cdots.$$

consider what happens when $r > 1$. The expression for the *finite* sum,

$$S_n = a + ar + ar^2 + \cdots + ar^n = \frac{a}{1-r} - \frac{a}{1-r}r^{n+1}$$

is still valid since $r \neq 1$. However, as n approaches infinity this expression itself becomes arbitrarily large. It does *not* approach any finite number. Thus the infinite geometric series does not have a "sum" in this case.

Nevertheless, when $r > 1$ it may still be of interest to compute S_n, the sum of finitely many terms. In this case it is convenient to rewrite S_n as

$$S_n = \frac{a}{r-1}(r^{n+1} - 1).$$

Example 4. Rework the chain letter question from Example 1 of Section 6.1.

Solution. The total number of people involved in 8 weeks will be

$$1 + 5 + 5^2 + 5^3 + \cdots + 5^8.$$

This is S_8 with $a = 1$ and $r = 5$. Thus

$$S_8 = \frac{a}{r-1}(r^{n+1} - 1) = \frac{1}{4}(5^9 - 1) = 488{,}281.$$

Next comes a physically motivated application of an infinite geometric series.

Imagine a ball bouncing on a hard surface. Its behavior can be described fairly well by saying that each time it hits the ground it bounces back to r times its previous height, where r is some number between 0 and 1 (see Figure 1). (The value of r depends on the particular ball and the surface on which it

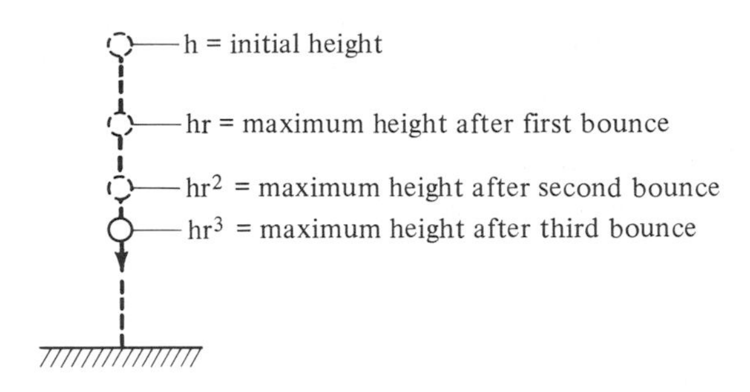

Figure 1.

bounces.) Thus, theoretically, the ball will bounce infinitely often—at lower and lower levels.

How far does it travel altogether?

Example 5. A ball is dropped from a height of 5 feet above a concrete sidewalk. Each time it hits the ground it bounces back to $\frac{2}{3}$ of its previous height. Find the total distance the ball travels.

Solution. After the initial fall of 5 feet, the ball bounces back up to a height of $5(\frac{2}{3})$ feet, and then falls that distance to the sidewalk again. So, by this time it has traveled a total of

$$5 + 5(\tfrac{2}{3}) + 5(\tfrac{2}{3}) = 5 + 10(\tfrac{2}{3}) \text{ feet.}$$

Now it bounces back to a height of $5(\frac{2}{3})^2$ feet and falls the same distance—an additional $10(\frac{2}{3})^2$ feet—for a total of

$$5 + 10(\tfrac{2}{3}) + 10(\tfrac{2}{3})^2 \text{ feet.}$$

Continuing indefinitely to "add" up the distances traveled between bounces, one gets

$$5 + 10(\tfrac{2}{3}) + 10(\tfrac{2}{3})^2 + 10(\tfrac{2}{3})^3 + \cdots.$$

This is 5 plus a geometric series with $a = 10(\frac{2}{3})$ and $r = \frac{2}{3}$. (The first term, 5, does not fit the pattern of the series because it represents a one-way fall, and every other term represents a round-trip bounce—up and back down.)

Thus the total distance traveled is

$$5 + \frac{a}{1-r} = 5 + \frac{20/3}{1-\frac{2}{3}} = 5 + 20 = 25 \text{ feet.}$$

Having discovered that this ball, bouncing infinitely often, actually travels only a finite distance, the next question is, "For how long does it bounce?" Most people, after accepting the idea of infinitely many bounces, would say the ball keeps on bouncing forever (theoretically).

Time of Fall. To treat this type of question you need to know that an object dropped from a height of h feet hits the ground in (approximately) $\sqrt{h}/4$

seconds; and an object which bounces up to a height of h feet gets there in $\sqrt{h}/4$ seconds.

This follows from the description of "Free Fall" in Section 5.2. But, if you have not studied that section, just accept the statement above. (The statement ignores air resistance and considers only the effect of gravity.)

Now consider a ball, initially dropped from a height of h feet, and returning at each bounce to r times its previous height.

The ball first hits the ground after $\sqrt{h}/4$ seconds.

Then it bounces back to a height of hr feet, which takes $\sqrt{hr}/4$ seconds, and falls to the ground in another $\sqrt{hr}/4$ seconds. The total time elapsed thus far is

$$\tfrac{1}{4}\sqrt{h} + \tfrac{2}{4}\sqrt{hr} \text{ seconds.}$$

The next bounce up to height hr^2 and back takes $\sqrt{hr^2}/4$ seconds each way. So the total time now becomes

$$\tfrac{1}{4}\sqrt{h} + \tfrac{2}{4}\sqrt{hr} + \tfrac{2}{4}\sqrt{hr^2} = \tfrac{1}{4}\sqrt{h} + \tfrac{1}{2}\sqrt{h}[\sqrt{r} + (\sqrt{r})^2].$$

Continuing in this manner, the total time for the infinitely many bounces is found to be

$$\begin{aligned} &\tfrac{1}{4}\sqrt{h} + \tfrac{2}{4}\sqrt{hr} + \tfrac{2}{4}\sqrt{hr^2} + \tfrac{2}{4}\sqrt{hr^3} + \cdots \\ &= \tfrac{1}{4}\sqrt{h} + \tfrac{1}{2}\sqrt{h}[\sqrt{r} + (\sqrt{r})^2 + (\sqrt{r})^3 + \cdots]. \end{aligned}$$

Since $0 < r < 1$, so also $0 < \sqrt{r} < 1$. Thus the quantity in brackets above is a geometric series which has a sum. In this case the ratio is $\sqrt{r}$ rather than r itself. So the total time for the infinitely many bounces is

$$\tfrac{1}{4}\sqrt{h} + \tfrac{1}{2}\sqrt{h}\frac{\sqrt{r}}{1 - \sqrt{r}} \text{ seconds.}$$

Putting both terms over the common denominator $1 - \sqrt{r}$ simplifies this to

$$\frac{1 + \sqrt{r}}{1 - \sqrt{r}}\frac{\sqrt{h}}{4} \text{ seconds.} \qquad (*)$$

Example 6. When (if ever) does the ball in Example 5 stop bouncing?

Solution. Substitute into expression (*) the values $h = 5$ and $r = \frac{2}{3}$. (This is most easily done with the aid of a calculator.) You should find about 5.5 seconds.

Experiment. This example invites experimental verification. Take any ball which bounces reasonably well, and determine the value of r by measuring the height of its return bounces. Then drop it from a known height h feet and measure, with the second hand of a watch, the time until it stops bouncing. Compare this measured time with the time you calculate from expression (*) using *your* values of h and r.

The agreement will not be perfect. You will actually find that the ball bounces somewhat longer than the time predicted by (*). This is because the simplified Time-of-Fall discussion above did not take air friction into account. Because of air resistance, the true time for an object to fall to the ground from height h feet is somewhat more than $\sqrt{h}/4$ seconds. Thus the total time of bouncing is somewhat more than that given by (*).

Problems

1. Compute the sum (if any) of the geometric series

$$4 + 4(\tfrac{1}{2}) + 4(\tfrac{1}{2})^2 + 4(\tfrac{1}{2})^3 + \cdots$$

   (a) by the *method* illustrated in Example 1, and
   (b) by use of the general formula (if applicable).

2. Compute the sum (if any) of the geometric series

$$1 + (\tfrac{4}{3}) + (\tfrac{4}{3})^2 + (\tfrac{4}{3})^3 + \cdots.$$

3. Convert the repeating decimal $x = 1.23\overline{123}$ to an equivalent fraction
   (a) by regarding it as a geometric series, and
   (b) by the method of Section 2.3.

4. Use the method of summing a geometric series to convert each of the following to an equivalent fraction in lowest terms:
   (a) $x = \overline{0.1}$ (b) $x = 0.\overline{9}$ (c) $x = 21.\overline{309}$.

5. Suppose you were offered a job for 1 year only with a salary determined as follows. You get 1 cent for the first week, 2 cents for the second week, 4 cents for the third week, 8 cents for the fourth week, and so on until the year is over. Would you take the job?

6. Using the data on Ampay dealers in Problem 3 of Section 6.1, find the *total* number of dealers after five completely successful levels of recruitment.

7. A ball is dropped from a height of 4 feet, and each time it hits the ground it bounces back to only $\frac{1}{4}$ of its previous height. Find (a) the total distance the ball travels and (b) the total time it bounces.

8. Using your own ball, discover the value of r for bounces on some suitable hard surface. Then drop the ball from a measured height, h feet, and determine how long it bounces. How does this observed time compare with the time you calculate from (*)? [Note. Some "superballs" bounce on a hard surface with $r = 0.8$ (or more).]

9. Find a simplified general expression for the total *distance* traveled by a ball dropped from height h if at each bounce it returns to a height r times its previous maximum. This means redo the calculations of Example 5 for the general case and find the sum of the resulting geometric series. Your final simplified answer will be an analog of expression (*).

* 10. Imagine that every January 1 you deposit \$1000 into a savings account which pays 10% interest per year compounded annually. Assuming no withdrawals and ignoring the effect of taxes, show that the balance after 20 years would exceed \$63,000.

CHAPTER 7

Areas and Volumes

Try to answer the following questions:

(a) If a 10-inch pizza costs \$3.00, what should a 15-inch pizza cost?
(b) If a flower pot 1 foot deep and 1 foot in diameter holds 5 gallons, what would be the capacity of a similar pot 2 feet deep and 2 feet in diameter?
(c) How much more water will flow through a $\frac{5}{8}$-inch garden hose than through a $\frac{1}{2}$-inch hose (assuming the same pressure and length)?
(d) How many dimes can be placed flat on the face of a half-dollar with none overlapping each other or extending off the edge of the half-dollar?

The "obvious" answers for most people are (a) \$4.50, (b) 10 gallons, (c) 25% more water, and (d) two or three dimes. And *each is wrong.*

After you have studied this chapter it should be clear why the "obvious answers" are wrong; and by then the correct answers should be obvious.

7.1. Areas

Any discussion of area begins with the rectangle. As you know, a rectangle which is a units by b units (Figure 1) has area

$$A = ab \text{ square units.}$$

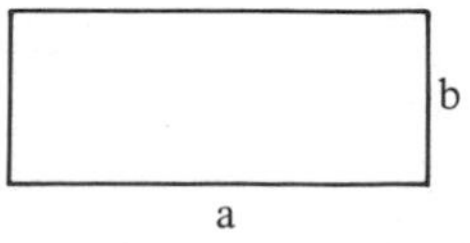

b

a

Figure 1.

Example 1. How does the area of a 4-foot by 6-foot rectangle compare with that of a 2-foot by 3-foot rectangle?

Solution. The smaller rectangle has area $2 \times 3 = 6$ square feet, and the larger one has area $4 \times 6 = 24$ square feet—four times that of the smaller.

This example illustrates the fact that if you double the dimensions of a rectangle, the area is quadrupled.

More generally one has the following.

Theorem. *If each dimension of a rectangle is multiplied by the positive number m, then the area is multiplied by m^2. (The letter m stands for "multiplier.")*

PROOF. If the original rectangle is a units by b units, it will have area $A = ab$. The new rectangle will then be ma units by mb units, which means its area is $ma \times mb = m^2ab = m^2A$. □

Example 2. If you have a square 10 inches on a side, what would be the dimensions of a square having half the area?

Solution. You must find the appropriate multiplier m so that $m^2 = \frac{1}{2}$. Thus $m = \sqrt{\frac{1}{2}} = 1/\sqrt{2} = \sqrt{2}/2$. So the new square should have side length

$$10m = 5\sqrt{2}$$

or approximately 7.07 inches.

What can one say about the areas of other plane regions, which are not rectangles?

If they are sufficiently simple, such as triangles or trapezoids, their areas can be computed by tricks which relate them to rectangles. See the figures on page 34, for example.

But, more generally, mathematicians prefer to *define* the area of a plane region by first approximating it with many small rectangles. Suppose you wanted to know the area of the region enclosed by a certain curve as in Figure

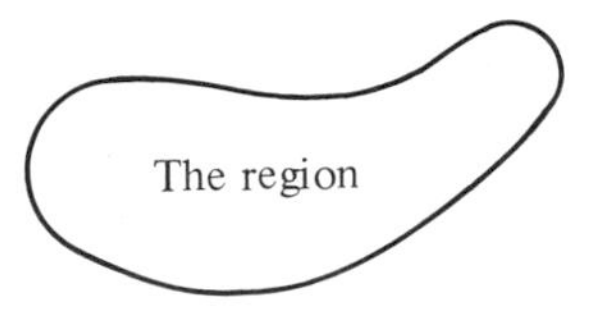

Figure 2.

2. You might first approximate it by considering a mesh or gridwork of horizontal and vertical lines which create many small rectangles as in Figure 3. The shaded rectangles in this figure are those which lie entirely inside the original region. The sum of their areas is an approximation to what you want

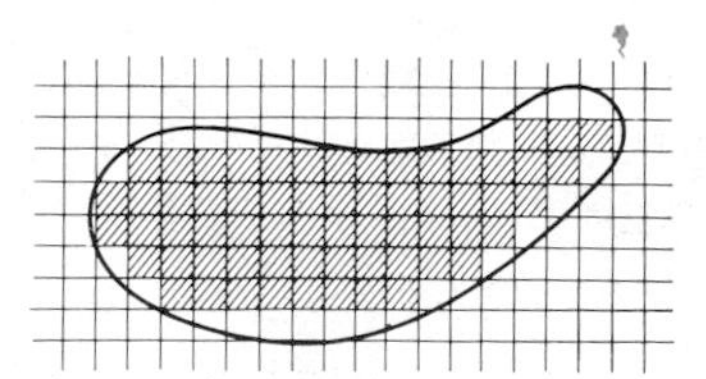

Figure 3.

to call the area of the original region. Moreover, the approximation would become better, in general, if you repeated the process using more closely spaced, and hence more numerous, mesh lines. Now imagine letting the mesh spacing, both vertical and horizontal, "approach zero." Then the various approximations—the sums of the areas of the rectangles lying entirely inside the original region—should approach what one would call the area of the original region.

All this can actually be made precise, but that requires some sophisticated mathematics. The above heuristic ideas will suffice to justify a useful generalization of the previous theorem on enlarging (or shrinking) a rectangle.

Theorem. *Let A be the area of the region in a plane enclosed by some curve—not necessarily a smooth curve. If this region is changed into a "similar" region by multiplying all distances (or dimensions) by a positive number m, then the new region will have area $= m^2 A$. (Two regions are said to be similar if one is like a photographic enlargement of the other.)*

PROOF. Let the original region be the one depicted in Figures 2 and 3. Then in Figure 4 draw a similar region obtained by multiplying all distances by m, and superimpose a mesh of horizontal and vertical lines *entirely similar* to those of Figure 3. Thus the mesh spacing is m times what it was in Figure 3. (In these figures m is about $\frac{3}{2}$.)

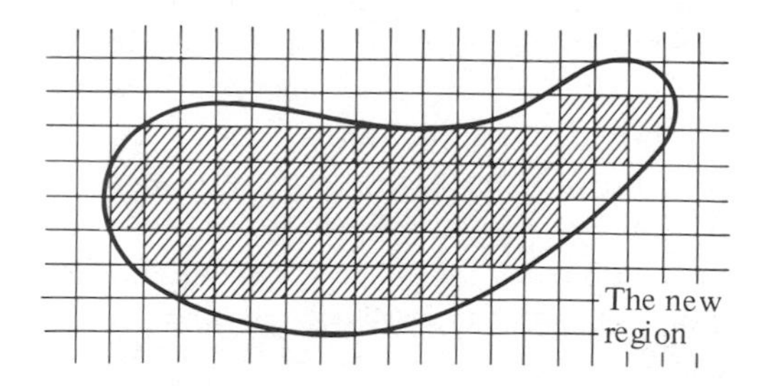

Figure 4.

Because the figures are similar, each of the shaded rectangles approximating the new region has area exactly m^2 times the area of its counterpart in the original region of Figure 3. Thus the total approximation for the new area is m^2 times that for the original area.

Finally imagine increasing the number of mesh lines while the mesh spacings all approach zero in both Figures 3 and 4 always preserving the similarity of the two figures. It then follows that the exact area of the new region is m^2 times that of the original. □

Example 3. If a 10-inch pizza costs \$3.00, what should a 15-inch pizza cost? Assume that both pizzas are round, of the same thickness, made from the same recipe, and that there is no discount for "quantity".

Solution. The circular disk for the larger pizza can be obtained from that for the smaller by multiplying all distances by $m = 15/10 = 3/2$. Thus the larger pizza has area equal to $(3/2)^2 = 9/4$ times that of the smaller. Since the thicknesses are assumed equal, the larger pizza will weigh 9/4 times the weight of the smaller. Thus it should cost \$3.00 × (9/4) = \$6.75.

How should the price of a pizza depend on its diameter?

If you have the slightest doubt about the validity of this conclusion, perform the following experiment. Cut two circular disks out of the same piece of cardboard, the larger having diameter equal to $\frac{3}{2}$ that of the smaller. Now compare the weights of the two disks with the aid of a small scale.

There is another way of handling Example 3, if you know that the area of a circle is πr^2, where r is the radius. Suppose you have two circles, one of radius r and the other of radius mr. (In Example 3 these radii would be 5-inches and $7\frac{1}{2}$-inches.) Then the first circle would have area πr^2 and the second would have area $\pi(mr)^2 = m^2(\pi r^2)$. Thus the area is multiplied by m^2—or in the case of Example 3 by $\frac{9}{4}$. The advantages of using the method presented on the previous pages are (1) it does not matter whether you know a formula for the area of a circle or not and (2) the underlying reason why areas are multiplied by m^2 becomes clear.

Another special type of region which could be analyzed without this section is the region inside a triangle. A triangle with base b and altitude h has area $\frac{1}{2}bh$. So a similar triangle with base mb and altitude mh has area $\frac{1}{2}(mb)(mh) = m^2(\frac{1}{2}bh)$. Again, this confirms that multiplying the dimensions by m multiplies the area by m^2.

Problems

1. (a) How many square inches are in a square foot?
   (b) How many square inches are in a square yard?
   (c) How many square feet are in a square yard?
   (d) How many square centimeters are in a square meter?
   (e) How many square meters are in a square kilometer?
   (f) How many square centimeters are in a square inch? (1 inch = 2.54 centimeters.)
2. If it takes 1 pound of grass seed for an area 10 feet by 20 feet, how much is required for an area 20 feet by 40 feet?
3. You are buying wall-to-wall carpet for two rooms, one 10 feet by 12 feet and the other 15 feet by 18 feet. If the carpet for the smaller room costs $120, what would you expect to pay for the same type of carpet for the larger room?
4. A 10-foot wall is covered with drapes at a cost of $150. What would you expect to pay to cover a 15-foot wall (in a room with the same height ceiling) with drapes of the same type?
5. If you want to plant a lawn in a triangular region with sides 60 feet, 80 feet, and 100 feet long, for how many square feet should you buy seed and fertilizer?
6. If a knit scarf 8 inches by 30 inches required 160 feet of yarn, how much would be needed for a comparable scarf 6 inches by 20 inches?
7. A rectangle 220 feet by 198 feet contains an area of 1 acre.
   (a) What would be the dimensions of a similar rectangle containing 3 acres?
   (b) What is the area (in acres) of a house lot measuring 110 feet by 99 feet?
   (c) How many acres are in a square mile?
8. If a 12-inch pizza costs $4.00 and a 16-inch pizza costs $6.50, both having the same thickness, which gives the most pizza per dollar?
9. Assume that you are opening a pizza restaurant, and your small pizzas will be 10 inches in diameter. To the nearest inch, what should be the diameter of your large pizzas if you want them to be twice the size (area) of the small ones?
10. Assume that a 15-inch pizza costs $5.50. What would an 18-inch pizza cost if it had the same thickness, and were made from the same recipe, and there were no quantity discount?
11. The amount of water (or blood) which flows through a pipe (or artery) of a given length is proportional to the cross sectional area of the pipe, and the difference between inlet and outlet pressures. Garden hoses are usually sold in two standard sizes—$\frac{1}{2}$-inch and $\frac{5}{8}$-inch—representing their inside diameters. Consider a $\frac{1}{2}$-inch hose and a $\frac{5}{8}$-inch hose, both of the same length, both lying horizontally, and both connected to water sources of the same pressure with their outlets open to the air. How much more water will flow through the larger hose in a given period of time?
12. A triangle with side lengths of 90 inches, 26 inches, and 80 inches has an area of 1008 square inches. What is the area of a triangle

(a) with side lengths 90 yards, 26 yards, and 80 yards?
(b) with side lengths 120 inches, 39 inches, and 135 inches?

13. Assume that it takes 3 gallons of paint to cover the walls of a room 8 feet by 9 feet. How many gallons would be required to paint the walls of a room 16 feet by 18 feet with the same height ceiling?

7.2. Volumes

Just as the discussion of areas begins with the rectangle, so the discussion of volumes begins with the rectangular solid. A rectangular solid which is a units by b units by c units (Figure 1) has volume

$$V = abc \text{ cubic units.}$$

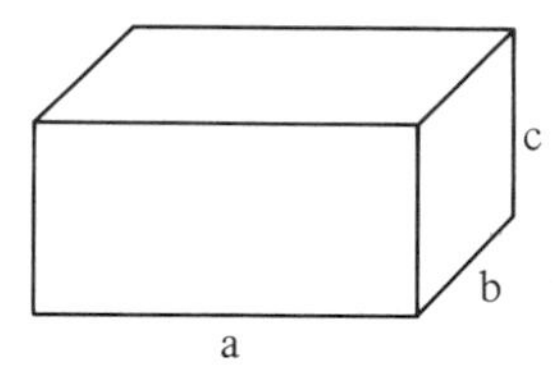

Figure 1.

Example 1. How many cubic inches are in a cubic foot?

Solution. A cubic foot can be represented by a rectangular solid each edge of which has length of 1 foot, or 12 inches. So the volume in cubic inches is $V = 12 \times 12 \times 12 = 1728$.

Let us next consider how the volume of a solid changes if one multiplies all dimensions by a positive constant m. The first simple observation is the analog for rectangular solids of the first theorem of Section 7.1 (for rectangles).

Theorem. *If each dimension of a rectangular solid is multiplied by the positive number m, then the volume is multiplied by m^3.*

PROOF. If the original rectangular solid is a units by b units by c units, it will have volume $V = abc$. The new rectangular solid will then be ma units by mb units by mc units, which means that its volume is

$$(ma) \times (mb) \times (mc) = m^3abc = m^3V. \qquad \square$$

Example 2. If a rectangular block of ice 6 inches by 1 foot by 1 foot weighs 31 pounds, what is the weight of a rectangular block of ice 1 foot by 2 feet by 2 feet?

Solution. The larger block has exactly twice the dimensions of the smaller. So, using $m = 2$, one concludes that the larger block has a volume of $m^3 = 8$ times that of the smaller, and hence it weighs $8 \times 31 = 248$ pounds.

The theorem shows how to relate the volumes of two "similar" *rectangular* solids if the dimensions of one are m times the dimensions of the other. Does the same effect, multiplication of the volume by m^3, also hold for "similar" solids which are not rectangular?

The answer is "yes," and the proof is analogous to that for areas in Section 7.1.

One must first decide how to *define* the volume of an arbitrary solid (three-dimensional) region such as that in Figure 2.

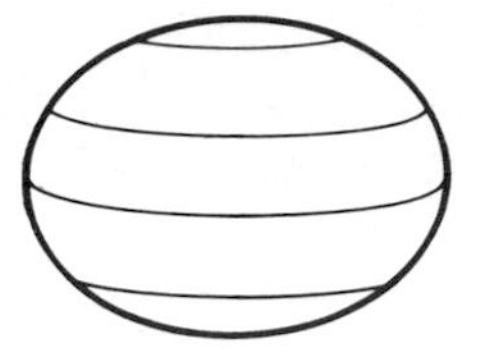

Figure 2.

By analogy to the definition of areas in Section 7.1, think of approximating the region with small rectangular regions—in this case rectangular *solids*. More specifically, imagine the region of Figure 2 sliced up by many horizontal and vertical *planes*. Some of the vertical planes are perpendicular to the page and some are parallel to the page. These will produce many small rectangular solids.

Figure 3 shows the rectangular solid produced by *just two* adjacent horizontal planes, *two* adjacent vertical planes perpendicular to the page, and *two* adjacent planes parallel to the page. One adds up the volumes of those rectangular solids which are *contained entirely in the original region* to get an approximation to the volume of the original region.

Finally, imagine introducing more and more planes parallel to the existing

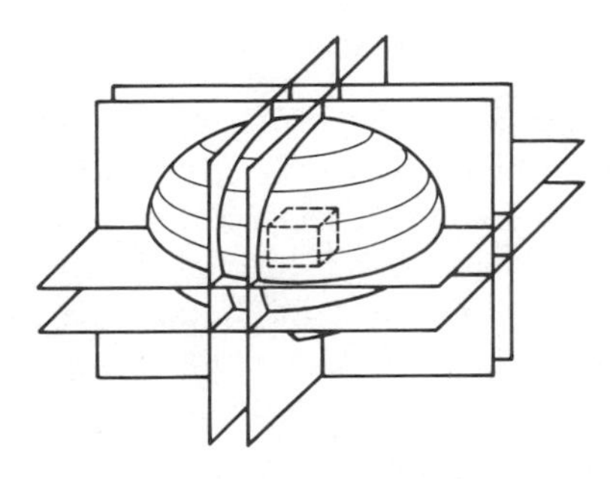

Figure 3.

ones while the separation between adjacent parallel planes (the mesh spacing) approaches zero. The resulting approximations—the sums of the volumes of those rectangular solids lying entirely inside the original region—approach what we will call the volume of the original region.

This concept of volume permits a proof of the following result relating "similar" solid regions. Two solid regions are said to be **similar** if one could be obtained from the other by some sort of imaginary three-dimensional photographic enlarging process. All distances within the region should be multiplied by the same positive factor m (the multiplier).

Theorem. *Let V be the volume of the region in space enclosed by some surface—not necessarily a smooth surface. If this region is changed into a similar region by multiplying all distances (or dimensions) by a positive number m, then the new region will have volume $= m^3 V$.*

PROOF. The proof is analogous to the one given for areas in Section 7.2. Visualize analogous constructions and arguments, now taking place in three-dimensional space.

Thus, consider two similar solid regions with the new one (Figure 4) having dimensions equal to m times those of the original region (Figure 3). On the original region superimpose a mesh of planes as described above, and on the new region an entirely *similar* mesh with spacing equal to m times that on the original. Figure 4, like Figure 3, displays just one of the many rectangular solids produced by the mesh of planes.

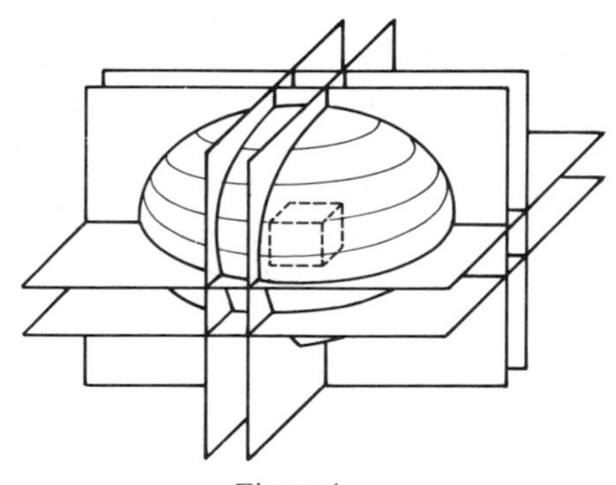

Figure 4.

The rectangular solid shown in Figure 4 corresponds to the one shown in Figure 3 under the similar plane mesh. The new one (in Figure 4) has m times the dimensions, and hence m^3 times the volume of the original rectangular solid (in Figure 3). The same relationship holds between each of the other rectangular solids (not shown) in Figure 3 and its counterpart in Figure 4. Adding up the volumes of those rectangular solids lying inside each region, one obtains an approximation to the volume of the new region which is exactly m^3 times that for the original region.

Finally, as the mesh spacings approach zero—always *preserving similarity* between the mesh on the original and the mesh on the new region—this ratio of m^3 between the approximations to volume is always the same. Thus the actual volume of the new region is m^3 times that of the original. □

Example 3. If a ball of string 2 inches in diameter contains 400 feet of string, what will be the content of a ball of the same type of string 3 inches in diameter? Assume that both balls are wound equally tightly.

Solution. The larger ball of string has a diameter equal to $\frac{3}{2}$ times that of the smaller. So $m = \frac{3}{2}$; and the larger ball has volume $m^3 = 27/8$ times that of the smaller one. It follows that the larger ball contains

$$\frac{27}{8} \times 400 = 1350 \text{ feet of string.}$$

There is another way of handling Example 3, if you know that the volume of a sphere is $\frac{4}{3}\pi r^3$, where r is the radius. Suppose you have two spheres, one of radius r and the other of radius mr. (In Example 3 these radii would be 1 inch and $\frac{3}{2}$ inch.) Then the first sphere would have volume $\frac{4}{3}\pi r^3$ and the second would have volume $\frac{4}{3}\pi(mr)^3 = m^3(\frac{4}{3}\pi r^3)$. Thus the volume is multiplied by m^3—or in the case of Example 3—by 27/8. The advantages of using the method presented on the previous pages are (a) it does not matter whether you know a formula for the volume of a sphere or not and (b) the underlying reason why volumes are multiplied by m^3 hopefully becomes clearer.

Many other special shaped regions can be analyzed without this section. Consider, for example, a cylinder with circular base of radius r and with height h (Figure 5). The volume of such a cylinder is $\pi r^2 h$. So a similar cylinder with

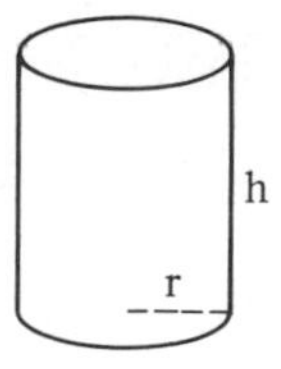

Figure 5.

base of radius mr and with height mh has volume $\pi(mr)^2(mh) = m^3(\pi r^2 h)$. Once again, multiplying the dimensions by m has the effect of multiplying the volume by m^3.

Experiment. Measure the height and the diameter of a standard 46-ounce juice can. Then do the same for a single-serving 6-ounce juice can. You will find that the dimensions of the large can are almost twice those of the small can. Thus the larger ought to contain almost $6m^3 = 6 \cdot 2^3 = 48$ ounces. And indeed 46 ounces is close.

A large juice can has dimensions about double those of a small can.

Example 4. How many dimes could you place flat on the face of a half-dollar with none overlapping each other or extending beyond the edge of the half-dollar? To answer this question you need to know that dimes, quarters, and half-dollars are essentially similar solids with volumes proportional to their values. (This dates back to the days before 1965 when dimes, quarters, and half-dollars were made of silver.)

Solution. Let the dimensions (radius and thickness) of a half-dollar be m times those of a dime. Then the volume of a half-dollar is m^3 times that of a dime. But this ratio of the volumes must be 5. So $m^3 = 5$. You do not need a table of cube roots to see that m must be less than 2. (Actually, $m = \sqrt[3]{5} = 1.71$, rounded off.) Thus the diameter of a half-dollar is less than twice that of a dime. So you could only lay *one* dime on top of a half-dollar without overlap.

As a practical matter, how could one determine the volume of a solid which is *not* rectangular, cylindrical, spherical, or some other simple shape?

This is the problem which faced the great mathematician Archimedes (ca. 287–212 B.C.) when King Hieron II of Syracuse wanted to know if his new crown was really made of gold. Perhaps Hieron had been cheated and the crown was basically silver covered by a thin layer of gold.

Now Archimedes could easily have distinguished between say a 1-inch cube of gold and a 1-inch cube of silver. Gold weighs about 1.8 times as much as an equivalent volume of silver. (One says gold is 1.8 times as "dense" as silver.) But how could he determine the volume of the crown so as to compare its weight with an equal volume of gold?

The answer came to Archimedes as he was getting into his bath tub one day. If the tub were filled to the brim with water, then the water which spilled over the edge as he immersed himself completely would have volume equal to his own body volume. (By catching and measuring the water that overflowed, he could determine the volume of an immersed object of any shape.) According to legend, Archimedes was so excited at this observation that he jumped out of the tub and ran down the street without his clothes shouting, "Eureka! I have found it."

In particular, Archimedes now had a method for finding the volume of Hieron's crown. And indeed the crown did not weigh as much as it would have had it really consisted of that much gold.

King Hieron II on a bronze coin of Syracuse, Sicily, 275–215 B.C.

PROBLEMS

1. (a) How many cubic feet are in a cubic yard?
   (b) How many cubic inches are in a cubic yard?
   (c) How many cubic centimeters are in a cubic meter?
   (d) How many cubic centimeters are in a cubic inch?

2. If a cylindrical ice cream tub 4.5 inches deep and 8 inches in diameter holds 1 gallon, what is the capacity of a cylindrical tub which is
   (a) 9 inches deep and 16 inches in diameter?
   (b) 9 inches deep and 8 inches in diameter?
   (c) 4.5 inches deep and 16 inches in diameter?

3. If a ball of yarn 3 inches in diameter weighs 2 ounces, what would be the weight of a similar ball of yarn 5 inches in diameter?

4. How many cubic yards of gravel are needed to cover a driveway 300 feet long and 9 feet wide with a layer 6 inches thick? [*Note*. Contractors and suppliers dealing with gravel, loam, sand, or concrete drop the word "cubic" and refer to so many "yards" (of gravel, loam, sand, or concrete). They really mean cubic yards.]

5. A "board foot" of lumber is a rectangular piece of wood 1 foot by 1 foot by 1 inch, or a piece of equivalent volume. How many board feet are in an 8-foot "two by four," i.e., a piece 2 inches by 4 inches by 8 feet long? (Lumber is usually priced by the board foot.)

6. If a gold coin 32 mm in diameter and 2 mm thick contains 1 ounce of gold, what is the gold content of a coin 24 mm in diameter and 1.5 mm thick made of the same alloy?

7. How many cubic yards of concrete are needed for a rectangular foundation 12 inches thick and 8 feet high for a house 24 feet by 40 feet (outside dimensions)?

8. A hot tub 5 feet in diameter requires 300 gallons of water to give a depth of 2 feet. How much water would be required for the same depth in a hot tub 6 feet in diameter?

9. The Federal Emergency Management Agency offers plans for making your basement into an "expedient fallout shelter" if nuclear war threatens. Here are excerpts: "Provide overhead barrier by placing 12 inches of earth on roof or on floor over basement ... Improve vertical barrier by placing earth against all exposed basement walls."

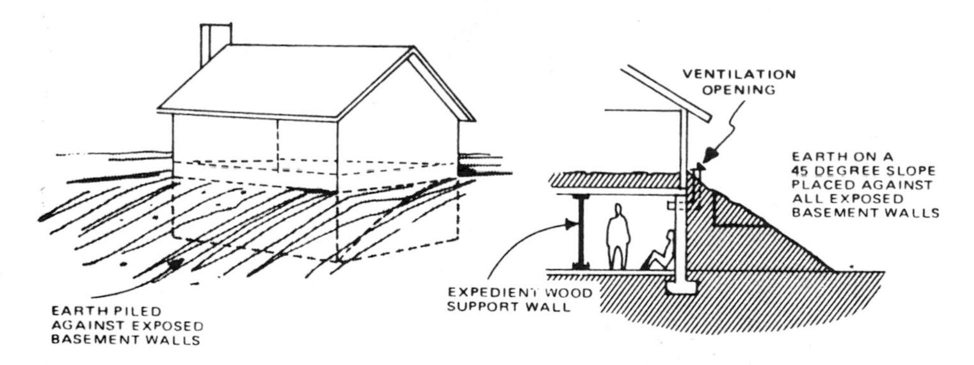

Figure 6.

For a *small* house 24 feet by 32 feet with a walkout basement, (a) How many wheelbarrows of dirt at $2\frac{1}{2}$ cubic feet each would be needed? (b) If you do not have a wheelbarrow, how many 2-gallon bucket loads would it take? (c) What would this dirt weigh at 100 pounds/cubic foot (assumed dry and free of stones)? (d) How long would the job take (assuming the ground is not frozen or rocky)?

10. Suppose you want to estimate the weight of a boulder about 6 feet by 5 feet by 4 feet. You look around for a small stone of the same material, and you find one which is about 4 inches by 5 inches by 3 inches. If the small stone weighs 2.5 kilograms, show that the boulder must weigh $9\frac{1}{2}$ tons. [*Note.* One kilogram is equivalent to about 2.2 pounds.]

11. Six cubic yards of black dirt are spread uniformly over an area of 2700 square feet for a future lawn. How thick is the resulting layer of black dirt?

12. Explain why it generally requires less water to boil two eggs than it does to boil one?

13. If the soil mixture for a 6-inch diameter flower pot includes 1 teaspoon of a certain fertilizer, how much of this fertilizer would you expect to put in the mixture for a similar 9-inch diameter pot?

7.3. Surface Area of a Solid (Versus Volume)

Section 7.1 dealt with areas of regions in a flat (two-dimensional) plane, and Section 7.2 dealt with volumes of three-dimensional regions.

Nothing has been said yet about areas of nonflat surfaces. However, there are many situations in which one needs to consider both the volume of a solid and the surface area of that solid.

Only for a few special types of surfaces can one exactly calculate the surface area. So, once again, the emphasis will be more on *comparisons* of similar regions.

As in Section 7.2, consider a given region enclosed by some surface in three-dimensional space (the original region); and compare this with another similar region in which all distances are m times those of the original region. To compare the total surface areas, divide up the original surface into many small "patches" as suggested by Figure 1, and divide up the surface of the new region into *similar* "patches." If these patches are sufficiently small, they can

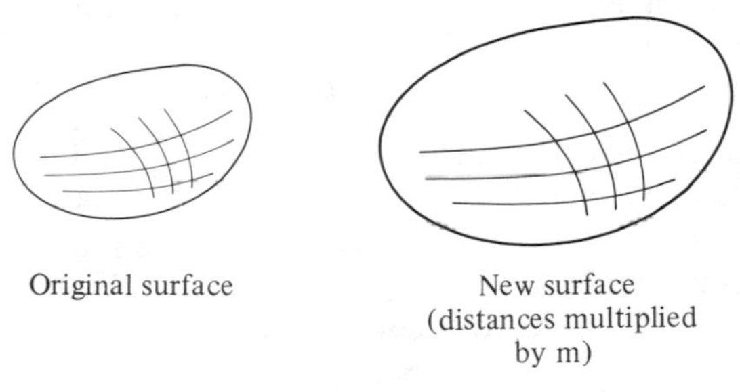

Figure 1.

be considered as approximately flat. Thus, by Section 7.1, each patch of the new surface has area equal to m^2 times that of its counterpart on the original surface. Adding up the areas of all patches on the two surfaces leads to the following conclusion.

Theorem. *Let A be the area of the surface enclosing a solid region. If this region is changed into a similar region by multiplying all distances (or dimensions) by a positive number m, then the new region will have surface area $= m^2 A$.*

Example 1. Confirm the assertion of the above theorem for the special case of a circular cylinder.

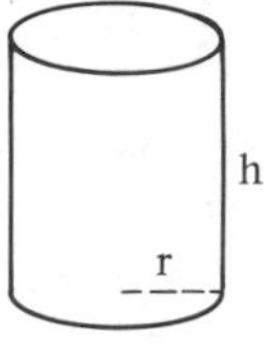

Figure 2.

Solution. Let the original cylinder have radius r and height h (see Figure 2). Then its total surface area can be computed as follows. The top and bottom of the cylinder are circles, each having area πr^2. The curved surface has height h and circumference $2\pi r$. One can imagine unrolling it to produce a flat rectangle h units by $2\pi r$ units. This rectangle will have area $2\pi rh$. So the total surface area for the cylinder is

$$A = 2\pi r^2 + 2\pi rh.$$

Now consider a similar cylinder with radius mr and height mh. By the same reasoning it will have a total surface area of

$$2\pi(mr)^2 + 2\pi(mr)(mh) = m^2(2\pi r^2 + 2\pi rh) = m^2 A.$$

Both the surface area and the volume of a solid are important in consideration of

(a) the rate of cooling of a heated object,
(b) the cost of canning beans,
(c) the rate of fall of an object through the air,
(d) the ability of certain insects to walk on water.

Consider, for example, the cooling of a heated object. The amount of heat in the object depends upon its volume and the type of material it is made of, and of course its temperature. For a given material at a given temperature, the heat content is proportional to the volume. On the other hand, the rate at which heat escapes (in given surroundings) is proportional to the surface area.

Example 2. If a dime and a half-dollar are both heated to 150°F and then placed apart from each other at room temperature, which will cool faster, and why?

Which will cool faster, and *why*?

Solution. The half-dollar is a solid which is similar to the dime and with all dimensions multiplied by $m = \sqrt[3]{5} = 1.71$. (See Example 4 of Section 7.2.) Thus the half-dollar initially contains $m^3 = 5$ times as much heat as the dime, but it has only $m^2 = (1.71)^2 \cong 3$ times the surface area. Thus the half-dollar will cool down more slowly than the dime.

The following example would be too difficult to assign as a problem at this stage. But hopefully, after reading it you will be able to analyze other questions involving both volumes and areas of given objects.

Example 3. The largest dinosaur was the Brachiosaurus—up to 80 feet long, weighing 50 tons, as tall as a three-story building. Is there any theoretical limit to the size of an animal?

Solution. Imagine two "similar" animals, one having m times the dimensions of the other where $m > 1$, but both built according to the same plan and using the same materials. Thus, in particular, their bones are similar and made of the same material. Then the strength of the bones will be proportional to their *cross-sectional* area. Thus the bones of the larger animal will have m^2 times the strength of the smaller. On the other hand, the larger animal will weigh m^3 times as much as the smaller.

Now, if you imagine the effect of larger and larger values of m, you note that the weight which must be supported by the bones increases faster than the strength of the bones. Thus, for a given pattern of construction and given materials, there will be some size beyond which the animal could not support its own weight.

PROBLEMS

1. Compute the surface area, A, of a rectangular solid a units by b units by c units. Then use your result to confirm that a similar rectangular solid with m times the dimensions will have surface area $= m^2A$.

2. The surface area of a sphere of radius r is known to be $A = 4\pi r^2$. Use this result to confirm that a sphere with m times the dimensions will have surface area $= m^2A$.

3. If it requires 55 square inches of tin to make a can which holds 1 pound of beans, how many square inches of tin are required for a *similar* restaurant-size can holding 8 pounds?

4. If an empty tin can designed to hold 6 ounces of tomato juice weighs $1\frac{1}{2}$ ounces, what would be the weight of a *similar* empty can designed to hold 48 ounces of tomato juice? (Assume both cans are made of the same material with the same thickness.)

5. In Problem 11 of Section 7.1 you found that a $\frac{5}{8}$-inch garden hose carries 56.25% more water than a $\frac{1}{2}$-inch hose (for a given length and pressure). Assume both hoses are made of the same type of vinyl and of the same thickness, and each is 50 feet long. (a) How much heavier will the $\frac{5}{8}$-inch hose be than the $\frac{1}{2}$-inch hose? Think. (b) How much more would the raw materials cost to make the $\frac{5}{8}$-inch hose than the $\frac{1}{2}$-inch hose?

6. Suppose you want to store 2000 gallons of solar-heated water, and you have a choice between buying two round cylindrical tanks holding 1000 gallons each, or one larger *similar* tank holding 2000 gallons. Which of these alternative storage arrangements will cause the water to cool down more slowly? *Explain.*

7. You have built a window display for a store using wax models draped with textiles. The wax cost \$12.50 and the textiles cost \$37.50. Your boss likes the display and asks you to reproduce it with double the dimensions. What will the materials cost for the larger display?

8. To maintain the temperature of a building in winter, one must supply heat to replace that which escapes through the walls, roof, and floor. And, as already noted, the rate at which heat escapes through a surface is proportional to the area of that surface (other things being equal).

   During one relatively mild winter a certain small house, 24 feet by 32 feet (by 12 feet high) was maintained at 65°F with a total fuel bill of \$600. Estimate the total fuel bill for maintaining an indoor temperature of 65°F in a block of *eight* apartments, 48 feet by 64 feet (by 24 feet high), as in Figure 3, in the same neighborhood during the same winter.

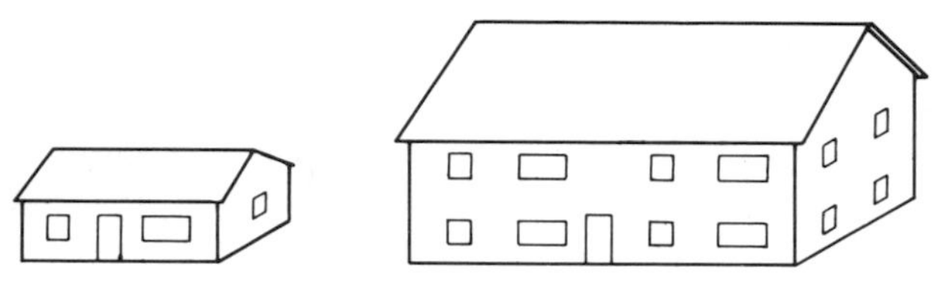

Figure 3.

9. If a person and an ant fall out of a fifth floor window, the person is killed but the ant walks away. Why?

10. Why can birds fly, while a person with comparable sized homemade wings cannot?

11. The earth is approximately a sphere of radius 4000 miles, and about 25% of the surface area is land (excluding Antarctica).
    (a) How many acres of land are there?
    (b) How much land is there per person on the average?
    (c) If the world population continued to grow at 1.8% per year, how much land would there be per person in the year 2400? (This presumes knowledge of Section 6.3.)

12. Why can certain insects walk on water, while people cannot? (This has to do with *surface tension*—the property which makes the surface of a body of water behave something like a stretched elastic membrane.)

7.4. Computation of Cube Roots

In some problems involving volumes, it is of interest to find the cube root of a given number.

For example, if a can which holds 8 pounds of beans is similar to a can which holds 1 pound of beans, then the larger can has dimensions m times those of the smaller where $m^3 = 8$. Here it is easy to see that $m = \sqrt[3]{8} = 2$.

But how would we find m if $m^3 = 2$, or 3, or 4, or 5?

Let us try to mimic the method for finding square roots described in Section 3.3.

To find $\sqrt[3]{a}$, where a is a given positive number, begin with a guess $x_1 > 0$. Now if x_1 were exactly $\sqrt[3]{a}$ then

$$\frac{a}{x_1^2}$$

would be exactly x_1 again. Can you see why?

On the other hand, if $x_1 < \sqrt[3]{a}$, then $a/x_1^2 > \sqrt[3]{a}$, and if $x_1 > \sqrt[3]{a}$, then $a/x_1^2 < \sqrt[3]{a}$. So we might try averaging x_1 and a/x_1^2 to get x_2 —hopefully a better approximation to the true value of $\sqrt[3]{a}$.

This process *will* work. That is, if continued, it will produce a sequence of numbers getting closer and closer to the true value of $\sqrt[3]{a}$. However, it is agonizingly slow.

It can be shown via calculus that the "convergence" to the answer is much faster if one uses an appropriate weighted average. Specifically, one should define

$$x_2 = \left(\frac{a}{x_1^2} + 2x_1\right)\frac{1}{3}.$$

Add "one part" of a/x_1^2 and "two parts" of x_1, and then divide by 3.

Now the process continues, constructing the next approximation as

$$x_3 = \left(\frac{a}{x_2^2} + 2x_2\right)\frac{1}{3},$$

and so on.

Example 1. Find $\sqrt[3]{5}$ correct to two decimal places.

Solution. As an initial guess, let $x_1 = 2$. Then define

$$x_2 = \left(\frac{5}{2^2} + 2 \cdot 2\right)\frac{1}{3} = 1.75$$

and

$$x_3 = \left[\frac{5}{(1.75)^2} + 2(1.75)\right]\frac{1}{3} = 1.71$$

and

$$x_4 = \left[\frac{5}{(1.71)^2} + 2(1.71)\right]\frac{1}{3} = 1.71.$$

This repetition indicates that the answer is 1.71 correct to two decimal places.

Here are some further applications of cube roots.

Example 2. A 2-gallon flower pot is 8.5 inches in diameter (at the top). Find the diameter of a similar 5-gallon pot.

A 5-gallon pot and a 2-gallon pot.

Solution. Let the dimensions of the 5-gallon pot be m times those of the 2-gallon pot. Then $m^3 = 5/2 = 2.5$. Thus $m = \sqrt[3]{2.5}$. Let the initial guess for this cube root be $x_1 = 1$. Then

$$x_2 = \left(\frac{2.5}{1^2} + 2\cdot 1\right)\frac{1}{3} = 1.5,$$

$$x_3 = \left[\frac{2.5}{(1.5)^2} + 2(1.5)\right]\frac{1}{3} = 1.37,$$

$$x_4 = \left[\frac{2.5}{(1.37)^2} + 2(1.37)\right]\frac{1}{3} = 1.357,$$

$$x_5 = \left[\frac{2.5}{(1.357)^2} + 2(1.357)\right]\frac{1}{3} = 1.3572.$$

So $m = 1.357$ correct to four figures.

It follows that the diameter of the larger pot is

$$8.5m = 8.5(1.357) = 11.5 \text{ inches.}$$

Example 3 (for those who have studied Section 6.3, Example 5). A zero coupon CD (certificate of deposit) bought today for \$772.20 will be worth \$1000 when it matures in 3 years. Find the effective annual rate of interest, assuming annual compounding.

Solution. Let I be the effective annual rate of interest in percent, and let $i = I/100$. Then a deposit of \$772.20 accumulating interest for 3 years gives a balance of

$$\$772.20(1 + i)^3.$$

This must be the maturity value of \$1000. Thus

$$772.20(1 + i)^3 = 1000.$$

This gives $(1 + i)^3 = 1000/772.20 = 1.295$. So

$$1 + i = \sqrt[3]{1.295} = 1.09.$$

(This cube root is calculated by the method described above.) Thus $i = 0.09$, for an effective annual interest rate of 9%.

Note. You can sometimes use known cube roots to find new ones with the aid of the facts that

$$\sqrt[3]{ab} = \sqrt[3]{a}\,\sqrt[3]{b} \quad \text{and} \quad \sqrt[3]{a/b} = \sqrt[3]{a}/\sqrt[3]{b}.$$

Try to write out proofs of these analogous to the proofs of the corresponding properties of square roots in Section 3.1.

Example 4. Using some results from the examples above,

$$\begin{aligned}
\sqrt[3]{5000} &= \sqrt[3]{5}\,\sqrt[3]{1000} = 1.71 \times 10 = 17.1,\\
\sqrt[3]{0.005} &= \sqrt[3]{5/1000} = \sqrt[3]{5}/\sqrt[3]{1000} = 1.71/10 = 0.171,\\
\sqrt[3]{12.5} &= \sqrt[3]{5}\,\sqrt[3]{2.5} = (1.71)(1.357) = 2.32.
\end{aligned}$$

Note that none of the results available thus far give any such quick solution for $\sqrt[3]{50}$ or $\sqrt[3]{500}$, since $\sqrt[3]{10}$ and $\sqrt[3]{100}$ are not immediately available.

PROBLEMS

1. Compute $\sqrt[3]{2}$ correct to two decimal places.
2. Compute $\sqrt[3]{3}$ correct to two decimal places.
3. Compute $\sqrt[3]{4}$ correct to two decimal places.
4. Try to compute $\sqrt[3]{5}$ correct to two decimal places by starting with the guess $x_1 = 2$ (as in Example 1) and then defining $x_2 = (5/x_1^2 + x_1)\frac{1}{2}$, and so on. This is the natural type of averaging, reminiscent of the method for finding square roots. You should find that the resulting sequence, $x_1, x_2, x_3, \ldots$ does approach $\sqrt[3]{5}$—but much more slowly than it did in Example 1.
5. If an empty can designed to hold 2 pounds of peas weighs 2 ounces, what would be the weight of a similar empty can designed to hold 8 pounds of peas? Assume both cans are made of the same material with the same thickness.
6. If a basket 12 inches deep and 16 inches in diameter holds a bushel of apples, what would be the dimensions of a similar basket which holds a peck of apples? [*Note*. One bushel = 4 pecks.]
7. If a ball of yarn 3 inches in diameter contains 300 feet of yarn, what would be the diameter of a 600-foot ball of the same type of yarn (wound equally as tight)?

8. Extend the reasoning in the solution of Example 4 of Section 7.2 to conclude that (a) (diameter of a half-dollar) ÷ (diameter of a quarter) = $\sqrt[3]{2}$, and (b) (diameter of a quarter) ÷ (diameter of a dime) = $\sqrt[3]{2.5}$. (c) Check these conclusions by measurement.

9. If a 2-gallon flower pot is 8.5 inches in diameter (at the top) what are the diameters of (a) a similar 1-gallon pot and (b) a similar 3-gallon pot?

10. A zero coupon CD which costs $1000 today will be worth $1,231.93 after 3 years. Find the effective annual rate of interest, assuming annual compounding.

11. A zero coupon CD which costs $1000 today matures to $2000 in 6 years. Find the effective annual rate of interest. [*Hint*. To solve $(1 + i)^6 = a$, note first that $(1 + i)^3 = \sqrt{a}$.]

12. Find quickly, in your head, (a) $\sqrt[3]{8000}$, (b) $\sqrt[3]{8{,}000{,}000}$, (c) $\sqrt[3]{0.027}$, (d) $\sqrt[3]{27/8}$, (e) $\sqrt[3]{0.064}$.

13. Use results from Example 1 and Problems 1, 2, and 3 to find (a) $\sqrt[3]{16}$, (b) $\sqrt[3]{32}$, (c) $\sqrt[3]{5/8}$, (d) $\sqrt[3]{1/9}$, (e) $\sqrt[3]{6}$, (f) $\sqrt[3]{10}$.

14. If n is any positive integer and a is any positive number, here is a procedure for finding $\sqrt[n]{a}$ (the nth root of a) analogous to those in Sections 3.3 and 7.4 for square roots and cube roots. Start with a reasonable guess x_1. Then define

$$x_2 = \left[\frac{a}{x_1^{n-1}} + (n-1)x_1\right]\frac{1}{n}, \qquad x_3 = \left[\frac{a}{x_2^{n-1}} + (n-1)x_2\right]\frac{1}{n},$$

and so on. Use this method to compute (a) $\sqrt[4]{3}$ and (b) $\sqrt[5]{2.2}$.

15. What size TV screen (diagonal measure to the nearest inch) would give three times the screen area of a 10-inch TV. (Think.)

CHAPTER 8

Galilean Relativity

This chapter considers a variety of problems involving relationships between moving objects—a plane flying in moving air, a horn sounding on a moving train or car, a police radar speed trap, and the basics of sailing.

Many such problems are best analyzed by considering the points of view of different observers. Thus, for example, a boat traveling in a river might be considered from the point of view of a swimmer treading water nearby or from the point of view of an observer standing on the bank of the river.

8.1. Displacement and Velocity Vectors

This section considers the problem of combining two or more displacements (moves), or two or more velocities.

If a passenger on a train walks 10 feet forward (toward the engine) while the train is traveling 500 feet forward, then from the point of view of an observer on the ground the passenger has moved $500 + 10 = 510$ feet forward.

Similarly, if a plane flies with air speed v in air which is moving in the same direction with speed w, then an observer on the ground would say the plane has speed $v + w$. One also says the plane has speed $v + w$ *relative to* the ground. If the same plane now turns around and flies against the wind then it will have speed $v - w$ relative to the ground. (Assume $v > w$.)

Notice that a displacement or velocity is not completely described unless both its *magnitude and direction* are specified. Thus when a plane has air speed v in a body of air moving with speed w, one cannot say whether the speed of the plane relative to the ground is $v + w$, or $v - w$, or something else unless the directions are specified.

If two displacements or velocities have exactly the same or exactly opposite directions, then simple addition or subtraction is appropriate. Other cases can be more complicated.

Example 1. A man rows a boat across a 40-foot wide river flowing east. He always keeps the boat pointed due north (perpendicular to the shore). When he reaches the other bank he discovers that he has been carried downstream 30 feet from his starting point (Figure 1). How far did he travel relative to the earth?

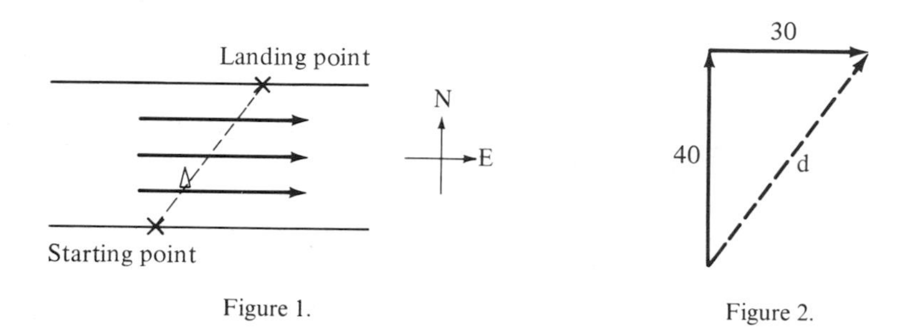

Figure 1. Figure 2.

Solution. His total travel or displacement can be considered to be composed of his displacement relative to the water, 40 feet north, plus the displacement of the water, 30 feet east (Figure 2). The net distance traveled, d, as determined by an observer on shore, is found with the aid of the Pythagorean Theorem

$$d^2 = 40^2 + 30^2 = 2500.$$

Thus

$$d = 50 \text{ feet}.$$

The lines with arrowheads used in Figure 2 to represent the two displacements are called "vectors." Their lengths and directions correspond to the amounts and directions of the displacements. The net displacement is also a "vector." In this example it has magnitude 50 feet and direction as shown by the dotted line.

In general a **vector** is a quantity described by a numerical magnitude and a direction. It is represented pictorially by an arrow in the appropriate direction and having length proportional to the magnitude of the vector.

Addition of two vectors (Figure 3) is accomplished geometrically by drawing the first vector, and then drawing the second with its tail placed at the tip of the first, as in Figure 4. The **sum of the vectors** is the new vector represented as a dotted arrow in Figure 4 extending from the tail of the first to the tip of the second original vector.

It is important to note that this definition is not affected by your choice of which vector to call the "first" and which to call the "second." Figure 5 shows

Figure 3.

Figure 4.

why. In Figure 5, the vectors of Figure 3 are added by the two possible methods—using the two possible choices of which vector is called the "first." Their sum (the dotted vector) is the same in each case—the diagonal of the resulting parallelogram.

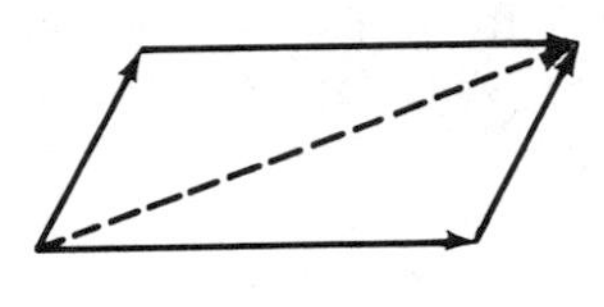

Figure 5.

Example 1 considered a special case in which the two given vectors had directions at right angles (90°) to each other. Other angles, such as in Figures 3 and 4, could be treated with the aid of a little "trigonometry"—not a subject for this book.

We shall usually restrict ourselves to the *simplest* special cases. Some examples in Section 8.3 will show how one can treat *certain* other angles besides right angles, still without the use of trigonometry.

Not only displacements, but also velocities are vectors, and can be added in the same way.

Example 2. A motorboat which travels 4 feet per second in still water is *pointed* north, directly across a river which has a current flowing to the east at 3 feet per second. What will be the actual speed of the boat as seen by an observer on the river bank?

Solution. Compare with Figures 1 and 2. Each second the boat moves 4 feet north and 3 feet east—a net distance of d where $d^2 = 3^2 + 4^2 = 25$. So $d = 5$ feet. Thus the speed relative to the bank is 5 feet per second. (*Speed* is the magnitude of velocity without regard to direction.)

Actually, there is no need to talk about the distance traveled by the boat in 1 second. One can work directly with the "velocity vectors" as suggested by Figure 6.

In practice, of course, one may prefer to cross the river directly, and not just accept whatever downstream drift the current imposes (as in Examples 1 and 2). In that case it is necessary to point the boat somewhat upstream to counteract the effect of the current.

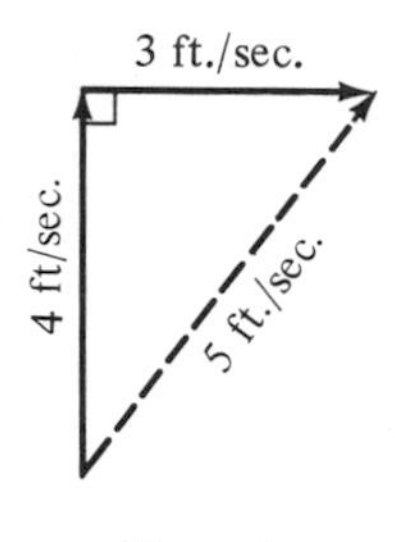

Figure 6.

Example 3. The same boat is crossing the same river as in Example 2, but now it is headed upstream at exactly the correct angle so that it will *actually move* directly across stream. What will be its speed relative to the bank?

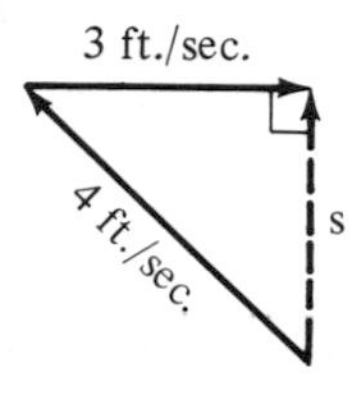

Figure 7.

Solution. The appropriate angle to point the boat is shown in Figure 7. The two velocity vectors to be added are shown as solid arrows, and their sum, the unknown velocity vector with magnitude s, is dotted. Once again the appearance of a right triangle permits use of the Pythagorean Theorem. This gives

$$3^2 + s^2 = 4^2.$$

So $s^2 = 16 - 9 = 7$, or $s = \sqrt{7} \cong 2.65$ feet per second.

Example 4. A boat with a speed of 4 feet per second relative to the water is to cross a river in which the current flows at 3 feet per second. If the river is 50 feet wide, (a) what is the minimum time required to reach the other side? (b) How long would the crossing take if one insisted on landing at the point directly opposite the starting point?

Solution. (a) To reach the other side as quickly as possible, keep the boat pointed perpendicular to the shore and do not worry about the fact that the current is carrying you downstream. (Figure 6) Then each second, you will move 4 feet closer to the opposite bank. So the crossing takes $50/4 = 12.5$ seconds.

(b) If you must reach the other side at the point directly opposite your starting point, then aim the boat as in Figure 7. Thus the speed relative to the bank is s where $3^2 + s^2 = 4^2$, or $s = \sqrt{16 - 9} = \sqrt{7} = 2.65$ feet per second. The crossing now takes $50/s = 50/2.65 = 18.9$ seconds.

PROBLEMS

[*Hint*. Draw diagrams.]

1. If you walk 2 miles east and then 1 mile north, how far will you be from your starting point?

2. If Amarillo is 275 miles east of Albuquerque ("as the crow flies") and Lubbock is 115 miles south of Amarillo, how far is Lubbock from Albuquerque?

3. Two children are playing marbles on a train traveling at 60 mph. A marble rolls along the floor of their railway car at a speed of 10 feet per second, first toward the front of the train and then toward the back. What would be the speeds of the marble in feet per second from the point of view of an "observer" on the ground outside the train?

4. A man rows a boat across a river 50 feet wide, always moving along a straight line between his starting point on one bank and the point on the opposite bank directly across the river. During the time it takes to make the crossing, the river flows 120 feet downstream. How far did the man row *relative to the water*?

5. A plane is flying at an air speed of 120 mph in a body of air moving south at 50 mph. Find the ground speed of the plane (i.e., its speed relative to the earth) if
   (a) the plane is pointed (and flying) south,
   (b) the plane is pointed (and flying) north,
   (c) the plane is pointed west (and flying in whatever direction this produces),
   (d) the plane is pointed in such a way that it actually flies due west relative to the earth.

6. The plane in Problem 5 flies to its destination d miles to the south and immediately returns. (a) How long does the round trip take? (b) How long would it have taken without any wind? (c) Which is less?

7. A motorboat traveling 10 feet per second relative to the water crosses a river 100 feet wide which flows east at 6 feet per second. The boat starts on the south bank and crosses to the north bank of the river. How long does the crossing take if
   (a) the operator keeps the boat pointed due north?
   (b) the operator points the boat in such a direction that it travels directly across the river to a point due north of its starting point?

8. A motorboat which travels 4 mph in still water went downstream a certain distance in a river having a current of 3 mph. The boat then returned to its starting point. If the round trip took 4 hours, how far downstream did the boat go?

*9. A plane with an air speed of v mph flies to its destination with a tailwind of w mph ($w < v$) and returns against a headwind of w mph. Find its *average* ground speed for the round trip. Is this more or less than v?

*10. The same plane as in Problem 9 must now fly to and from its destination to the east, while contending with a southerly wind (a cross wind) of w mph. What is its average speed for the round trip? Is this more or less than v?

*11. Two school teachers and a small boy left their campsite and walked 2 miles due south. Then they turned and walked 2 miles east, at which point they saw a bear. They hurried back to their camp by going 2 miles due north. What color was the bear? *Prove it.*

8.2. Doppler Effect

You have probably, at some time or another, noticed the changing sound of a train horn or a car horn as the vehicle approaches you, passes you, and departs. The sound changes from a high frequency (or pitch) to a lower frequency.

But an observer riding in the train or car hears no change at all.

This is one of the most striking and readily observed illustrations of the importance of considering the points of view of different observers.

To understand this phenomenon, one must first know that sound is propagated through the air as "waves." These are not waves of material moving up and down as in the ocean. Rather they are successive zones of higher and lower air pressure traveling away from the sound source. The source might be a piano note, or an electric horn, or any other device which produces alternately higher and lower air pressures in quick succession—specifically at a frequency which is detectable by the human ear.

In Figure 1 the sequence of successive high- and low-pressure zones in the sound produced by a factory whistle is suggested by the closer and wider spacing of concentric arcs emanating from the whistle. If the whistle is emitting a "pure tone" of a single frequency, then a rather simple pattern of alternately high- and low-pressure zones will just keep repeating itself.

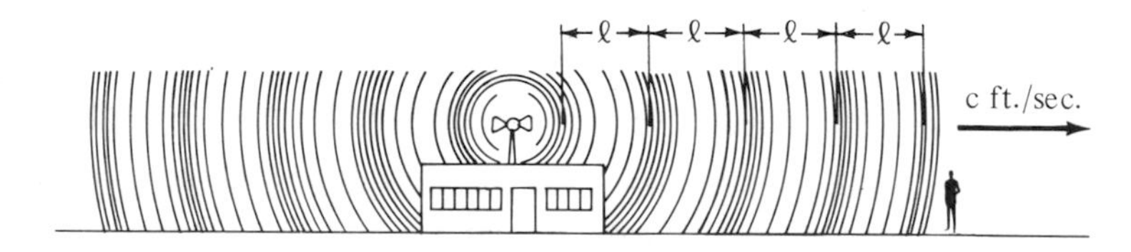

Figure 1.

The **frequency**, f, of the tone produced by the whistle is the number of complete cycles (from highest pressure to lowest to highest again) per second. Thus $1/f$ is the time in seconds between two consecutive high-pressure zones.

The assertion that "sound propagates like waves" means that the pattern of successive high- and low-pressure zones emanating from the whistle travels through the air, essentially unchanged, at a certain **speed of propagation**, c. Under normal circumstances for sound waves

$$c \cong 1100 \text{ feet/second (750 mph).}$$

In the time, $1/f$, between the production of two consecutive "highs" of pressure, the pattern will advance a distance $c(1/f)$, i.e., speed $\times$ time. Thus this is the distance in space between two consecutive highs. Hence it is the length in space of the repeating pattern of high- and low-pressure zones. This is called the **wavelength**, l, and we have the important relation

$$l = \frac{c}{f},$$

or

$$f = \frac{c}{l}.$$

Note that high frequencies correspond to short wavelengths and low frequencies correspond to long wavelengths.

Example 1. The lowest "C" on the piano corresponds to a frequency of about 32.7 cycles per second. What wavelength does this produce?

Solution. Taking $f = 32.7$ cycles per second and $c = 1100$ feet per second, we find

$$l = \frac{c}{f} = \frac{1100}{32.7} \cong 34 \text{ feet.}$$

The principal purpose of this section is to explain what happens when either the source of sound or the observer is moving.

Moving Observer. Consider first the case in which the source is stationary and the observer is moving—either toward or away from the source as in Figure 2. Once again, assume that the source produces a pure tone of frequency f cycles per second and wavelength $l = c/f$.

Suppose the observer is traveling *toward* the source at a constant speed v. Then he or she has an effective speed relative to the pattern of highs and lows of pressure of $c + v$.

Figure 2.

So this observer detects the pressure cycles coming more frequently than f cycles per second. In fact, since the wavelength is still l, the observer approaching the source detects one complete cycle (from high pressure to the next consecutive high) in

$$\text{time} = \frac{\text{distance}}{\text{speed}} = \frac{l}{c + v} \quad \text{seconds.}$$

Thus, he or she hears a new frequency,

$$f' = \frac{c + v}{l} = \frac{c + v}{c} f = \left(1 + \frac{v}{c}\right) f \quad \text{cycles/second.}$$

This is higher than the "true" frequency.

Similarly, an observer moving *away from* the source detects the cycles of high and low pressure coming less frequently than f per second. If this observer is traveling at speed v ($v < c$) away from the source, then the pressure zones are overtaking him or her at a speed $c - v$. Thus the time between consecutive high-pressure zones as detected by this observer is

$$\frac{l}{c - v} \text{ seconds,}$$

and that corresponds to a frequency

$$f'' = \frac{c - v}{l} = \frac{c - v}{c} f = \left(1 - \frac{v}{c}\right) f \quad \text{cycles/second.}$$

Example 2. Assume that the horn of a parked car is producing a tone of 300 cycles per second. What frequency is heard by an observer approaching this parked car at 45 mph and by an observer traveling away at 45 mph?

Solution. To convert the speed in miles per hour into feet per second, one must multiply by 5280 (the number of feet in a mile) and divide by 3600 (the number of seconds in an hour). Thus 45 mph corresponds to

$$45 \times \frac{5280}{3600} = 45 \times \frac{22}{15} = 66 \text{ feet/second.}$$

So the observer approaching the source hears frequency

$$f' = \left(1 + \frac{v}{c}\right) f = \left(1 + \frac{66}{1100}\right) 300 = 318 \text{ cycles/second,}$$

and the observer moving away hears

$$f'' = \left(1 - \frac{v}{c}\right) f = \left(1 - \frac{66}{1100}\right) 300 = 282 \text{ cycles/second.}$$

Moving Source. Now consider the case of a moving source and a stationary observer. The source might be the horn on a locomotive traveling at a constant speed v. Here the argument is a little different from the case of a stationary source and a moving observer, but the result is similar.

The moving source is actually producing a signal of a shorter wavelength ahead of its position, and of longer wavelength behind. Figure 3 should make this clear.

Let the frequency of the horn be f cycles per second, as heard by the crew in the locomotive. Consider a pulse of high pressure produced by the horn. Then the next "high" is produced $1/f$ seconds later. By that time the first pulse has traveled a distance $c(1/f) = c/f$, and the locomotive has traveled a distance $v(1/f) = v/f$. So ahead of the locomotive the distance between consecutive zones of high pressure is reduced to $c/f - v/f$, and behind the locomotive this

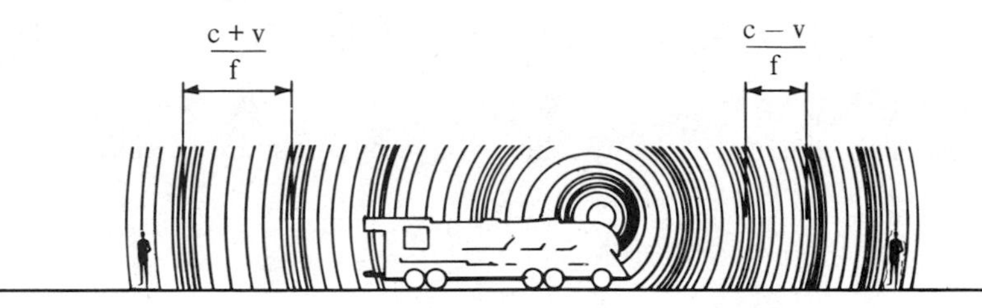

Figure 3.

distance is increased to $c/f + v/f$. That is, the wavelength becomes $(c - v)/f$ ahead of the locomotive and $(c + v)/f$ behind.

The frequency heard by an observer ahead of the source is, therefore,

$$f' = \frac{c}{\text{wavelength}} = \frac{c}{(c - v)/f} = \frac{f}{1 - v/c},$$

which is greater than f. And the frequency heard behind the source is

$$f'' = \frac{c}{\text{wavelength}} = \frac{c}{(c + v)/f} = \frac{f}{1 + v/c},$$

which is less than f.

In most day-to-day situations the speed v of the vehicle is small compared to the speed c of sound. And then it can be shown that good *approximations* are

$$f' \cong \left(1 + \frac{v}{c}\right)f$$

for the frequency heard ahead of the source, and

$$f'' \cong \left(1 - \frac{v}{c}\right)f$$

for the frequency heard behind the source. (See Problem 10.)

These approximate expressions should be easier to remember because they happen to be the same as the expressions found previously in the case of a moving observer and a fixed source.

Note that if a moving source approaches a stationary observer, passes him or her, and then continues to move away at the same speed v, the sound changes from a frequency of $(1 + v/c)f$ to $(1 - v/c)f$, approximately. But this change will not be sudden (unless the source passes directly through the observer). Rather, there will be a smooth transition from high to lower frequency as the source goes by. (This will be discussed in Section 8.3.)

Example 3. Assume that a car is approaching you with its horn blowing. As it comes toward you, the horn sounds like the note "E" above middle "C"—namely, 329.6 cycles per second. After the car passes you, the horn sound changes to middle "C"—namely, 261.6 cycles per second. How fast is the car going?

Solution. Assuming the car is traveling at a constant speed v, one has

$$\left(1 + \frac{v}{c}\right)f \cong 329.6,$$

$$\left(1 - \frac{v}{c}\right)f \cong 261.6.$$

Addition of these two equations yields $2f \cong 591.2$ or $f \cong 295.6$ cycles per second. Subtraction gives $2(v/c)f \cong 68$. So

$$2\left(\frac{v}{1100}\right)295.6 \cong 68.$$

Thus $v \cong 126.5$ feet per second. Converting this into miles per hour, one has

$$v \cong 126.5 \times \frac{3600}{5280} \cong 86 \text{ mph}.$$

Thus far there has been no explanation of the title of this section—Doppler effect.

Christian Johann Doppler (1803–1853) was an Austrian physicist. In 1842 he published a paper predicting that the color (frequency) of *light* from a moving source would depend on the velocity of the source, just as the pitch (frequency) of sound depends on the speed of its source. The moving sources which Doppler had in mind were moving stars. His result was approximately correct provided the speed of the source is much less than the speed of light—186,000 miles per second. Since then, the general phenomena of change in frequency of sound or light due to a moving source or a moving observer has become known as the Doppler effect.

Light, as well as sound, can be regarded as a wave phenomena (although this was not clear in 1842). However, light waves cannot be mentally visualized as well as water waves or sound (pressure) waves. Light waves propagate through space with no need for the presence of water, air, or any other material to carry them.

Without attempting to explain the electromagnetic theory of light waves, let us simply adopt the same expressions for frequency shift as for sound—those on page 131. The difference is that when dealing with light (instead of sound) $c = 186{,}000$ *miles per second* (instead of 1100 feet per second).

Perhaps the most familiar everyday application of the Doppler effect for electromagnetic waves is its use in radar speed detectors. These instruments send out a radio wave at a frequency f, and the reflected wave from a moving vehicle is received back at the radar set and compared with the original. Radio waves, like light, propagate at the speed $c = 186{,}000$ miles per second. The target vehicle acts as both a moving receiver and a moving source (of the reflected signal). This has the effect of doubling the change in frequency of the reflected wave. Thus, if the radar beam is reflected by a vehicle *approaching* the stationary radar set as speed v, then the returning wave will have frequency approximately

$$f' = \left(1 + 2\frac{v}{c}\right)f.$$

And if it is reflected from a *departing* vehicle traveling at speed v, the returning wave will have frequency approximately

$$f'' = \left(1 - 2\frac{v}{c}\right)f.$$

Example 4. A police radar signal reflected by an approaching car returns at a frequency $2 \times 10^{-5}\%$ higher than the transmitted frequency. Find the speed of the car.

Solution. If f is the original frequency transmitted, then the returning frequency is

$$f' = (1 + 2 \times 10^{-7})f.$$

Thus

$$\left(1 + 2\frac{v}{c}\right)f = (1 + 2 \times 10^{-7})f.$$

Therefore

$$2\frac{v}{c} = 2 \times 10^{-7}$$

or

$$v = 10^{-7}c = 0.0186 \text{ miles/second}.$$

In miles per hour this becomes

$$v = 0.0186 \times 3600 = 67 \text{ mph}.$$

In the problems which follow, use $c = 1100$ feet per second as the speed of propagation for sound waves. And use $c = 186{,}000$ miles per second or $c = 3 \times 10^8$ meters per second as the speed of propagation for light waves and radio waves.

A frequency of f cycles per second is sometimes referred to as f *hertz*. In other words 1 hertz is 1 cycle per second.

Problems

1. If you see a flash of lightning, and then hear the associated thunder 10 seconds later, how far away did the lightning strike?

2. In October and November 1980, the space probe Voyager 1 sent back remarkable pictures and other data from Saturn at a time when Saturn was one billion miles from earth. How long did it take for each signal from Voyager 1 to get back to earth?

3. A light year is the distance traveled by light in 1 year. How far is a light year in miles?

4. From one "C" note on the piano to the next higher "C" note is a distance of "one octave" and represents a doubling of frequency. For example, the lowest "C" on a piano has a frequency of 32.7 cycles per second, and "middle C" is three octaves higher. Thus "middle C" has frequency $= 2^3 \times 32.7 = 261.6$ cycles/second. The highest "C" on a piano is seven octaves above the lowest. Find its frequency and wavelength.

5. Radio waves travel at the same speed as light waves. Find the wavelength in meters of
   (a) an AM broadcasting station with a frequency of 1000 kilohertz,
   (b) an FM broadcasting station with a frequency of 100 megahertz.
   [*Note.* 1 kilohertz is 1000 hertz, and 1 megahertz is 10^6 hertz.]

6. Find the frequency of
   (a) violet light with a wavelength of 4×10^{-7} meters,
   (b) deep red light with a wavelength of 7.5×10^{-7} meters.

7. A locomotive horn produces a tone of 200 cycles per second. If the train is traveling at 60 mph, what frequencies are heard by an observer on the ground as the train approaches and after the locomotive has passed?

8. Astronomers have found that the spectra from distant nebulae (that is the characteristic light frequencies produced by various elements) is shifted toward lower frequencies by about 3.33% as compared to the same elements on earth. What does this "red shift" imply about the velocities of those nebulae?

9. If a police radar speed detector is targeted on a vehicle driving away at 100 mph, how will the frequency of the reflected radar signal differ from that transmitted by the radar set?

10. If the speed v of a moving vehicle is small compared to c, then

$$\left(1 + \frac{v}{c}\right)\left(1 - \frac{v}{c}\right) = 1 - \left(\frac{v}{c}\right)^2 \cong 1.$$

Therefore,

$$\frac{1}{1 - v/c} \cong 1 + \frac{v}{c}.$$

Use this to obtain the approximate expression for f' for the case of a moving source and a stationary observer. Similarly, obtain the approximate expression for f''.

* 11. (a) Rework Example 3 using the exact expressions, $f' = f/(1 - v/c)$ and $f'' = f/(1 + v/c)$, instead of the approximate ones. (You will find the *same* answer for v as obtained previously using the approximations.)
(b) Show that this is not just a coincidence. More specifically, show that for any observed values of f' and f'' you get $v/c = (f' - f'')/(f' + f'')$ regardless of whether you use the exact or the approximate expressions. (The computed values of f would *not* be the same.)

8.3. Components of Vectors

Section 8.1 offered several examples and problems involving addition of given vectors. Conversely, it is often useful to represent a *given* vector as the sum of two other vectors, conveniently chosen.

There are always infinitely many ways of choosing two vectors so that their sum is a given vector. In Figure 1, for example, the given vector v can be regarded as the sum of vectors x and y, or the sum of vectors u and w, or any number of other possibilities.

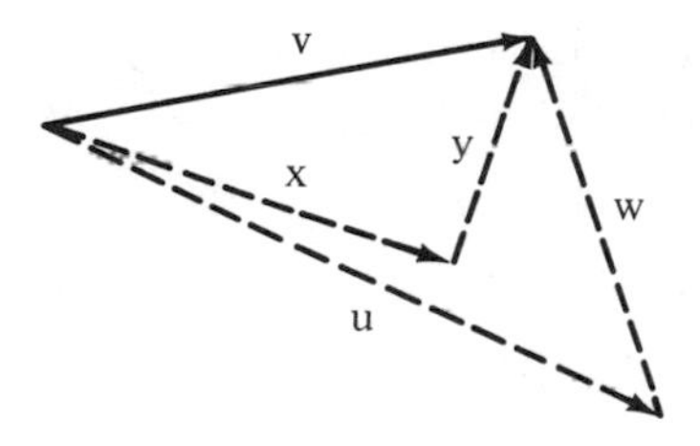

Figure 1.

One seeks a representation (or decomposition) of the given vector which will be useful for the problem at hand. The new artificial vectors whose sum is the original vector are called **components** of the original vector. Components are usually chosen at right angles to each other, such as x and y in Figure 1.

Example 1. A pedestrian walks east along the sidewalk for 40 feet, and then jaywalks across the road to the northeast (at an angle of 45°), reaching the other side after 30 feet (see Figure 2). How far is she from her starting point?

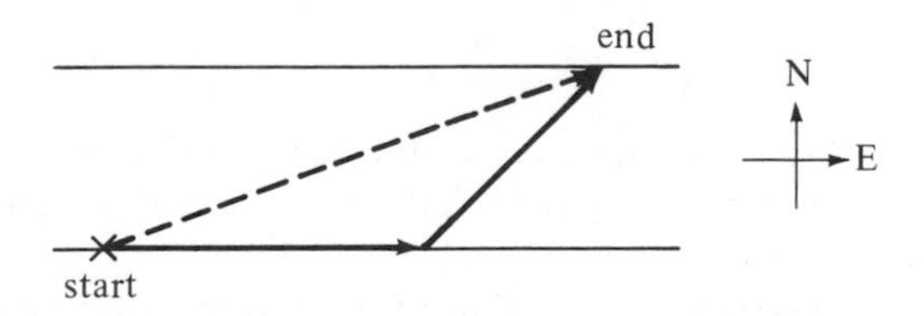

Figure 2.

Solution. The net displacement is the sum of the vectors shown in Figure 3. To find their sum it is convenient to regard the northeast vector of length 30 feet as being composed of two vectors (dotted in Figure 3), one pointing east and the other pointing north.

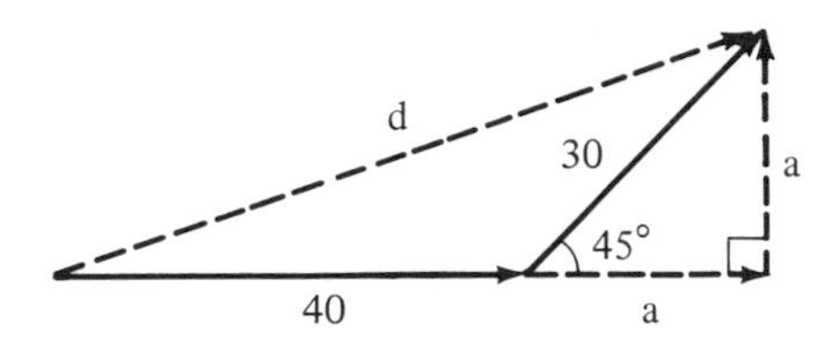

Figure 3.

Since the resulting triangle is an isosceles right triangle, these two new vectors have the same length, say a. Moreover, by the Pythagorean Theorem, $a^2 + a^2 = 30^2$. So $a^2 = 30^2/2$, or

$$a = \frac{30}{\sqrt{2}} = 15\sqrt{2}.$$

Now the problem of finding the length d is simply that of finding the hypotenuse of a new right triangle having legs of lengths $40 + 15\sqrt{2}$ and $15\sqrt{2}$. So

$$\begin{aligned} d^2 &= (40 + 15\sqrt{2})^2 + (15\sqrt{2})^2 \\ &= 1600 + 1200\sqrt{2} + 450 + 450 \simeq 4197. \end{aligned}$$

Thus

$$d \simeq 64.8 \text{ feet}.$$

Example 2. If you walk 1 mile east and then 2 miles northwest, how far are you from your starting point?

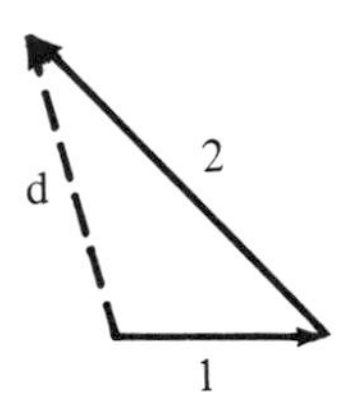

Figure 4.

Solution. One must add together the vectors shown in Figure 4. To do this, it is convenient to decompose the 2-mile northwest vector into a component west and a component north.

Again the resulting triangle (see Figure 5) is an isosceles right triangle. Thus the components have equal magnitudes, say a miles west and a miles north. So,

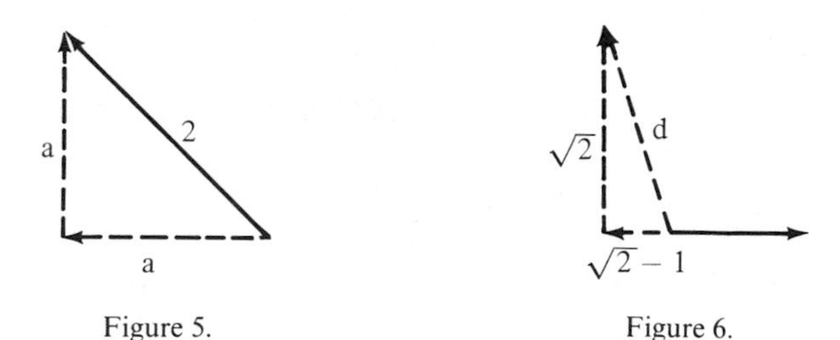

Figure 5. Figure 6.

by the Pythagorean Theorem, $a^2 + a^2 = 2^2$. Thus $a^2 = 2$, or $a = \sqrt{2}$ miles. Now the original problem can be treated by adding three vectors: 1 mile east, $\sqrt{2}$ miles west, and $\sqrt{2}$ miles north as in Figure 6. The net effect is to add $\sqrt{2} - 1$ miles west and $\sqrt{2}$ miles north. Thus

$$d^2 = (\sqrt{2} - 1)^2 + (\sqrt{2})^2 = 2 - 2\sqrt{2} + 1 + 2 = 5 - 2\sqrt{2} = 2.172.$$

Therefore

$$d = \sqrt{2.172} = 1.47 \text{ miles.}$$

Now reconsider Example 4 of Section 8.1.

Example 3. A motorboat which travels 4 feet per second relative to the water is crossing a river 50 feet wide in which the current flows at 3 feet per second. If the boat is headed partially upstream at exactly the correct angle so that it actually moves directly across the river, how long will the crossing take?

Solution. Whatever the angle of the boat, decompose its velocity into a component parallel to the banks and a component at right angles to this. Now the idea is to choose the angle of the boat so that the component of its velocity parallel to the bank is exactly 3 feet per second *upstream*. A little thought should convince you that this can be done in only one way. See Figure 7.

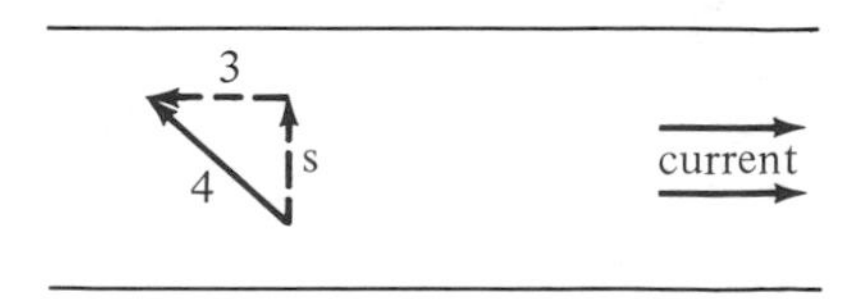

Figure 7.

Thus the velocity of the boat *relative to the water* is represented as the sum of two components—one is 3 feet per second upstream to counteract the current, and the other is across the stream with magnitude s. From the Pythagorean Theorem, $s^2 + 3^2 = 4^2$. So the speed across the stream (relative to the water or to the banks) is

$$s = \sqrt{16 - 9} = \sqrt{7} \text{ feet/second.}$$

And the time required to make the crossing is

$$\frac{50}{\sqrt{7}} = \frac{50\sqrt{7}}{7} \cong 19 \text{ seconds.}$$

There would be little point in introducing the idea of vector components if nothing came of it but new solutions of old problems. The main purpose of this section is to show how the decomposition of a vector into components can help to explain physical phenomena.

Example 4. The study of Doppler effect in Section 8.2 concentrated on observers and/or sources moving directly toward or directly away from each other. What does an observer actually hear as a locomotive passes him with its horn blowing (provided the observer is intelligent enough not to be standing directly on the tracks)?

Solution. Assume the locomotive is traveling at a constant speed v feet per second. The velocity vector of this moving source should be decomposed *at each instant* into a component toward the observer and a component at right angles to the first. It is only the component toward (or away from) the observer which is involved in the Doppler effect—and the components thus defined are continually changing, even though the velocity of the source is constant. See Figure 8. When the train is approaching from a distance

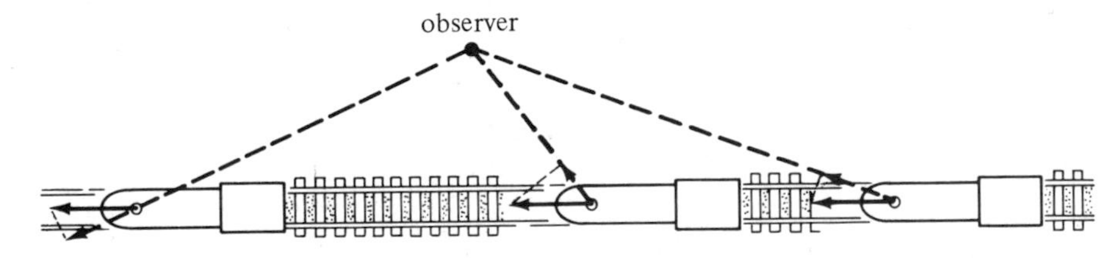

Figure 8.

(sounding a horn of frequency f) the observer does hear approximately the frequency

$$f' = \left(1 + \frac{v}{c}\right)f.$$

But as the locomotive gets closer to the observer, the component of v toward the observer gets smaller. Thus the Doppler increase in frequency *decreases* continuously as the locomotive passes the observer, and it becomes a decrease in frequency as the locomotive departs. Eventually the observer hears approximately the frequency

$$f'' = \left(1 - \frac{v}{c}\right)f.$$

Thus far we have considered two kinds of vectors—displacement vectors and velocity vectors. A third physical concept which has vector properties is force.

A force (like a displacement or a velocity) is specified by giving both its magnitude and direction. Moreover, forces which act at a given point can be added like vectors, and they can be decomposed into components.

A nice application arises in the theory of sailing. If you have never done any sailing, you might wonder why all the sailboats on a lake do not eventually end up on the downwind shore, helpless to move. How can one sail against the wind?

Sailboats travel in different directions using the same wind.

Consider a rudimentary sailboat with a single sail. Then the easiest thing to understand is sailing downwind. Given a south wind as in Figure 9, one can sail south be simply letting the sail out at right angles to the axis of the boat. Then the wind applies a force on the sail which moves the boat south.

In order to travel in any direction other than the direction the wind is

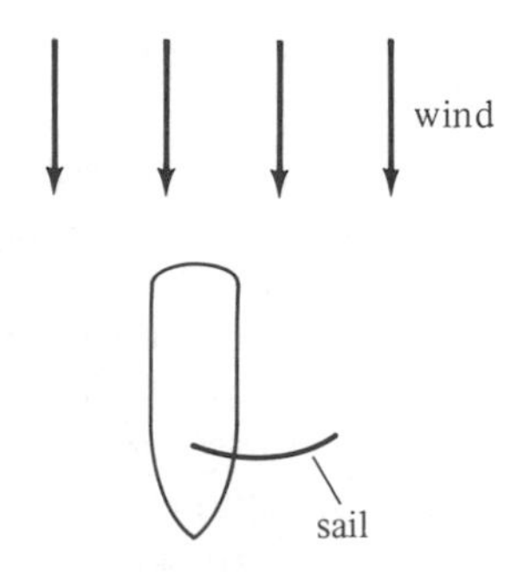

Figure 9.

blowing, the sailboat must have either a "keel" or a "centerboard"—something to resist sideways motion in the water.

Example 5. How can a sailboat sail at right angles to the wind direction?

Explanation. Let the wind be blowing south at a constant speed, and consider the sail position as shown in Figure 10. Then the wind produces a force on the sail which is approximately at right angles to the sail. That force F is decomposed into a component in the direction of the axis of the boat and a component at right angles to it. The latter component is cancelled out by the lateral force of the water on the keel (or centerboard). Thus the boat moves forward, across the wind.

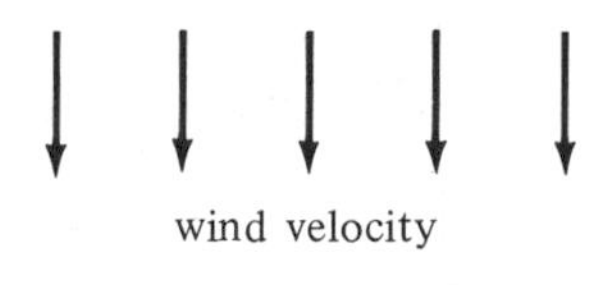

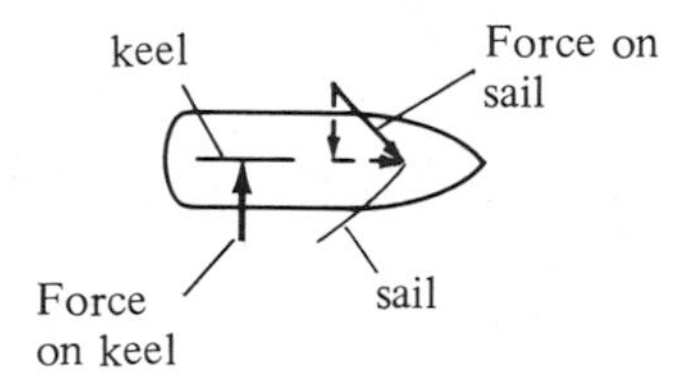

Figure 10.

As a matter of fact, a sailboat can move across wind much faster than the speed of the wind itself. (This may require angles for the sail other than that shown in Figure 10.) If one is going downwind, however, there is no way to sail faster than the speed of the wind.

Example 6. How can a sailboat sail upwind?

Explanation. It can't—at least not directly. However, it can "tack". If the wind is blowing south, for example, the sailboat can travel northeast for a while and then travel northwest, so that it is truly progressing north. The explanation is indicated in Figure 11. Again, when the sail is held in the proper position the wind applies a force on it approximately perpendicular to the sail. And again, what is of principal interest is the component in the forward direction. The other component, perpendicular to the axis of the boat, is counterbalanced by the lateral force on the keel.

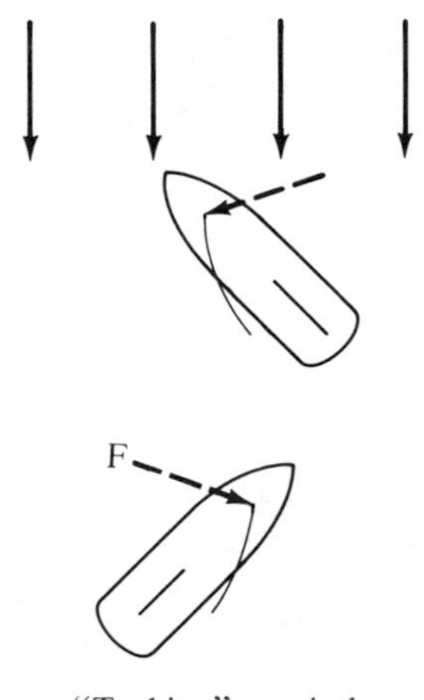

"Tacking" upwind

Figure 11.

Problems

1. If Philadelphia is 80 miles southwest of New York City and Albany is 130 miles north of New York City, how far is Albany from Philadelphia? (All distances are direct, of course.)

2. A plane with an air speed of 200 mph is flying in air which moves east at 40 mph. If the plane is pointed *northeast*, what is its ground speed?

3. A plane with an air speed of 130 mph is flying due east (relative to the ground) even though it has to contend with a wind blowing southwest at 70.7 mph. What is its ground speed?

4. A plane with an air speed of 100 mph flies from city A to city B 100 miles to the east. If the pilot has to contend with a steady southerly wind at 40 mph during the entire flight, how long does the trip take?

5. A police radar speed detector is not usually located directly in front of or directly behind the target vehicle. Will this cause the instrument to give too high or too low an indication of the target vehicle's speed? Explain.

6. A car passes you with its horn blowing. (a) What do you hear? (b) What does the driver of the car hear?

7. A bomb dropped from a plane whistles as it falls to earth. (Its speed and hence the frequency of the whistle are limited by air resistance.) (a) If a bomb were falling toward a point 2000 yards from you, what would you hear? (b) If it were falling directly toward you, what would you hear?

8. On a still day two small sailboats on a wide river are drifting downstream with the current. Both boats have their sails furled because there is no wind. But after a while, one of the sailors notices that there *is* a "wind"—at least he *thinks* he feels a

wind due to the forward motion of his boat. So this sailor decides to raise his sail and use the "wind" to tack downstream even faster than the current (he hopes).

Can he succeed in moving downstream faster than the other boat which still has its sail furled? Explain.

"I love hearing that lonesome wail of the train whistle as the frequency of the wave changes due to the Doppler effect."

CHAPTER 9

Special Relativity

The theory of relativity, due primarily to H. A. Lorentz in the 1890s and Albert Einstein in the early 1900s, revolutionized scientific thinking. This chapter gives a brief introduction to the ideas of "special relativity"—the subject of a 1905 paper by Einstein. The special theory of relativity considers the viewpoints of different observers moving with constant velocities, as in the previous chapter. But the key postulate and the conclusions will now be quite different and, for most people, surprising.

9.1. Simultaneity and Einstein's Postulate

Any study of physical phenomena requires the concepts of time and distance, and the ability to measure these quantities. The theory of relativity begins with a study of time and distance themselves—dramatically changing our understanding of these concepts.

Consider the problem of two different observers measuring time at two different places. Each has his own clock, and they would like to be sure that the two clocks are synchronized.

What should they do?

One possibility is for the two observers to bring their clocks together, and set them to read the same time. Then they could carry the clocks to the desired locations. But how could they be sure that nothing affects the clocks during transit?

Another obvious way to try to synchronize the clocks is for one observer to look at his clock and look simultaneously at the other observer's clock, using a telescope if necessary. If observer A sees that his clock reads the same as observer B's clock, he might say the two are synchronized.

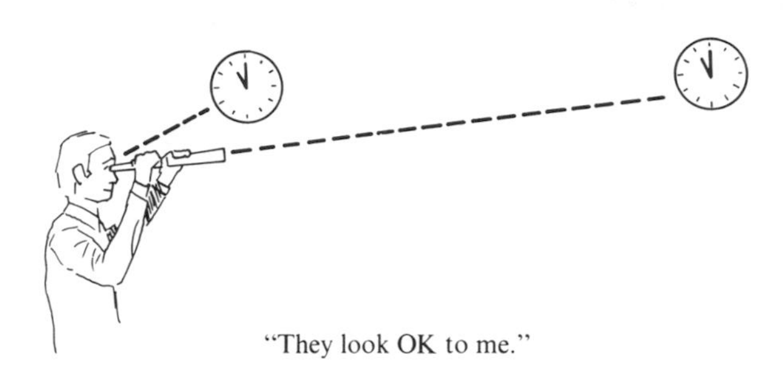

"They look OK to me."

But would observer B agree?

If A and B are distance d apart, and both are stationary, then the light which A observed from B's clock required time d/c to reach A, where c is the speed of light. Thus, if A simultaneously *saw* both clocks reading time t, then A *should* conclude that B's clock is actually fast by the amount d/c.

Under these circumstances, what would B see as the reading on A's clock when B's clock reads time t? (Problem 1.)

Because light does not travel instantaneously, the above procedure is not satisfactory for synchronizing two clocks at different places. Moreover, we know of no faster way to send signals. Electrical pulses, radio or radar signals, and light all travel at the same speed $c = 186{,}000$ miles per second.

But there is a simple remedy.

Clock Synchronization by Two Stationary Observers. Let the two observers carefully measure the distance, d, between their positions before they try to synchronize the clocks. And let them agree in advance that B will send a light signal when her clock reads time t. (This is entirely equivalent to A squinting through a telescope to look at B's clock and see it read t.) When A sees the light signal from B, he should make sure that his clock reads time $t + d/c$.

With this procedure, both observers will agree that the clocks are synchronized.

Another way of stating that the two clocks are synchronized at time t is to say that the "events" that A's clock reads t and B's clock reads t are simultaneous events.

The next question is, how can two observers decide whether or not two events are simultaneous if one observer is "stationary," and the other is moving?

Since we know of no faster way to communicate than via light signals, this question leads us to inquire a little more deeply about the propagation of light.

When light was first conceived as being a wave phenomenon it was naturally assumed that there must be some medium to carry the waves—just as air carries sound waves. However, it was also known that light could travel without the presence of air or any other known medium. So, as early as the seventeenth century, a medium called the "ether" was postulated as the basis for the propagation of light waves.

By the end of the nineteenth century, the existence of an "ether" was generally accepted.

If light waves propagated through the "ether" the way sound waves propagate through air, then one would assume that the speed of light is constant relative to the "ether," just as the speed of sound is approximately constant relative to the air.

But the best experiments all failed to detect the mysterious "ether."

In his 1905 paper, Einstein rejected the idea of "ether" and put forward the following remarkable idea.

Postulate. *Light from whatever source propagates with the speed c (approximately* 186,000 *miles per second) relative to any observer who can be considered "stationary" or any observer moving with constant velocity relative to a stationary one.*

Among the consequences of this postulate is the conclusion (not proved here) that there is no way—and never will be—of moving objects faster than the speed of light.

Note that the postulate says that light is *not analogous to sound,* inasmuch as there is *no* preferred type of observer (say stationary) who observes velocity c, while moving observers see another speed. Moreover, since the speed of light is said to be independent of the velocity of the source, light from a moving lamp is *not analogous to a ball thrown from a moving train.* One does *not* add the velocity of light to the velocity of the source.

The idea that the velocity of light is *independent of the velocity of its source* had little or no experimental basis in 1905. A debate over this point between Einstein and W. Ritz ensued in the scientific literature.

Later observations of "double stars" and experiments with elementary particles apparently confirmed Einstein's postulate.

Now consider the problem which two observers—one stationary and the other moving at a constant velocity—would have in trying to determine whether two events are simultaneous.

Imagine the following experiment. Observer A stands beside the railroad tracks watching a train of flatcars go by at speed v. Observer B is riding in the middle of the train carrying a flash gun. At each end of the train is a "slave" flash gun designed to respond to a light signal by instantly flashing its own light (Figure 1). Just as B passes A's position she flashes a light to signal the slave flash guns. As soon as the light signal reaches them, they each flash in response.

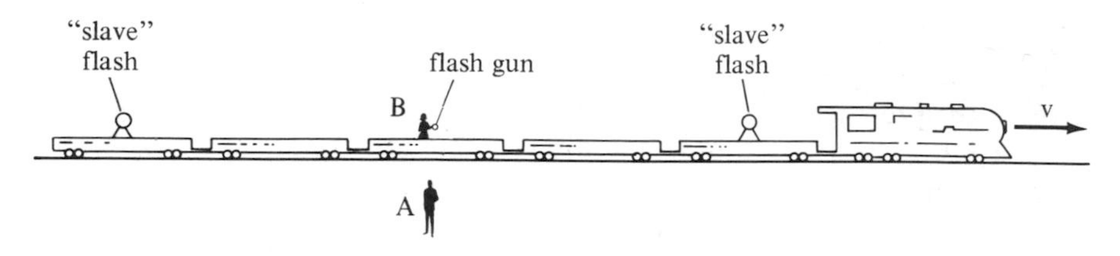

Figure 1.

Do the two slave flashes fire simultaneously?

Observer B sees that the answer is yes! Her light signal traveled at speed c, *relative to the train*, an equal distance to the two slave flashes. And she *saw* them fire simultaneously since their light signals propagated back to her at the same speed c, again over equal distances.

On the other hand, A sees B's light signal propagating forward and backward at speed c, *relative to the ground*. And at the same time he sees the train moving forward with speed v. So, as far as A is concerned, the slave flash at the rear of the train gets the light signal and responds to it *before* the one at the front.

Who is right, A or B?

According to Einstein, both are right. Two events which are simultaneous to one observer will not, in general, be simultaneous to another observer who is moving relative to the first.

Example 1. If, for the train in Figure 1, the distance between the slave flashes is $2d$, how much difference is there between the times of their firings according to observer A? Specifically, if $2d = 0.93$ miles and $v = 134$ mph (a long, fast train), what is the time difference?

Solution. The light signal from B travels at speed c relative to the ground and the train is moving at speed v. Thus, observer A reasons that the signal reaches the slave flash at the rear of the train in time t_1 where $ct_1 = d - vt_1$, i.e., in time

$$t_1 = \frac{d}{c+v}.$$

Similarly, the signal reaches the slave flash at the front in the longer time t_2 where $ct_2 = d + vt_2$, i.e., in time

$$t_2 = \frac{d}{c-v}.$$

The difference is

$$t_2 - t_1 = \frac{d}{c-v} - \frac{d}{c+v} = \frac{2dv}{c^2 - v^2}.$$

Now set $2d = 0.93$ miles and $v = 134/3600 = 0.0372$ miles per second. Notice that the denominator, $c^2 - v^2$, will be almost indistinguishable from

c^2. Thus

$$\frac{2dv}{c^2 - v^2} \cong \frac{2dv}{c^2} = \frac{(0.93)(0.0372)}{(186{,}000)^2} = \frac{0.93(3.72 \times 10^{-2})}{(1.86)^2 \times 10^{10}} = 10^{-12} \text{ second}$$

—one-trillionth of a second.

Problems

1. In the "experiment" described at the beginning of this section, observer A simultaneously sees both his own clock and B's reading time t, even though B's clock is distance d away. Under these circumstances, what would B see as the reading on A's clock when B's clock reads time t?

2. The moon is about 239,000 miles from earth. (Assume it is stationary.) If an astronaut on the moon reports that his clock reads 12 noon, what time should your clock read in order that the two clocks be synchronized?

3. The outermost planet in our solar system is Pluto, which is about 3.67×10^9 miles from the sun. The closest star (apart from the sun itself) is Proxima Centauri, which is about $4\frac{1}{4}$ light years away. (See Problem 3 of Section 8.2.) How many times the "radius" of our solar system is the distance to this star?

4. The galaxy Hydra is about four billion light years away. If our sun was formed five billion years ago, could intelligent life (if any) in the galaxy Hydra yet know that our sun exists?

5. In March 1981, astronomers at Lick Observatory of the University of California reported the discovery of four galaxies about 10 billion light years away. And these galaxies appeared to have been about six billion years old.
   (a) If our sun was formed five billion years ago, could intelligent life (if any) in these galaxies yet know that our sun exists?
   (b) If the universe started with a "big bang," how long ago (at least) did this occur?

6. A professional photographer uses a slave flash (which fires when it "sees" the main flash gun at the camera go off) to help light parts of the subject. If the slave flash is 20 feet from the main flash gun, (a) how long will it take to "see" the main flash? (b) Would the shutter of the camera still be open then?

7. An observer standing on the ground near the tracks sends a light-flash signal to the "engineer" just as the train's caboose passes the observer. The train is 0.93 miles long and is traveling at 144 mph. How long does it take before the "engineer" sees the light flash (a) as the engineer understands it? (b) according to the observer's calculations?

8. The space vehicle Voyager 1 was launched from earth in 1977. The trajectory had been preplanned for Voyager 1 to rendezvous with Saturn (in its orbit about the sun) and Saturn's moons (in their orbits about Saturn) more than 3 years after launch, and one billion miles from earth. Small corrections to Voyager 1's trajectory could be made from earth, but these had to be planned long in advance. If one tried to direct the space ship from earth while it was collecting data, (a) how long would a round-trip radio signal take? (b) at 50,000 mph, how far would Voyager 1 travel during this time?

9.2. Time Dilation

The previous section discussed the difficulty of synchronizing two clocks at time t, or more precisely, the potential conflict when two observers try to decide whether or not two events are simultaneous.

Now we want to consider how two clocks *run* if one is stationary and the other is moving at a constant velocity.

In the spirit of relativity—whether Galilean relativity or special relativity—there is no reason to refer to any observer as "stationary." Sometimes one chooses to refer to anything which is at rest relative to the "fixed stars" as being stationary. But, as far as physics is concerned, it would be equally valid to instead regard an observer moving with "constant velocity" as being "stationary."

A classroom or laboratory on earth is not really stationary since the earth is rotating on its axis and orbiting around the sun. Nevertheless, it will suffice for our purposes to consider an observer on earth as stationary.

Now, for each observer, whether stationary or moving with constant velocity, imagine a system of position markers and clocks distributed throughout space *at rest relative to this observer*. These are the tools that permit the observer to determine the position and time of any event in space by his or her standards. This collection of position markers and clocks is sometimes called the **reference frame** of the observer. He or she would insist that all the clocks in that reference frame be synchronized.

But what kind of clocks should one use?

There are cheap spring-driven pocket watches and high-quality wall clocks driven by weights and controlled by pendulums. There are electric clocks synchronized with the AC power supplies which feed them and there are battery operated clocks and watches controlled by quartz crystals. And the finest scientific instruments for measuring time are based on atomic standards.

Indeed the standard definition of the **second**, adopted in 1967 by the General Conference on Weights and Measures, is

> 9,192,631,770 periods of the radiation corresponding to the transition between the two hyperfine levels of the ground state of the atom Caesium 133.

Since the speed of light is constant, one can also imagine constructing a precise clock as follows:

Attach a light source and a mirror at either end of a rigid rod (or tube) of length d. Then let the basic unit of time be the time which elapses between a brief flash of the light source and the return of its reflection from the mirror back to the source. Such an **Einstein–Langevin clock** (see Figure 1) should be an acceptable time standard whether for stationary observers or for observers moving at a constant velocity.

But how does a stationary observer regard the clock of a moving observer?

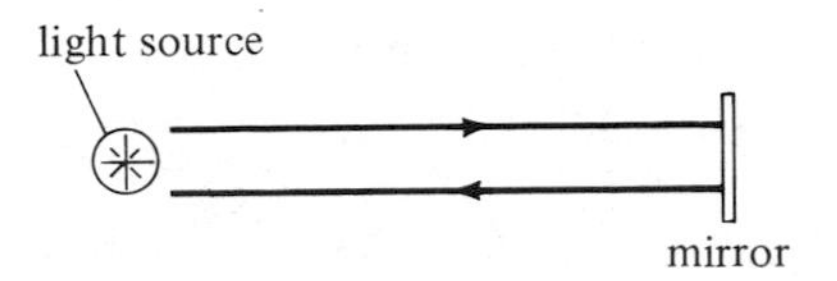

Figure 1.

The sketches in Figure 2 show an Einstein–Langevin clock as visualized by its owner who happens to be traveling in a space ship, and as interpreted by a stationary observer watching the space ship go by at velocity v. The clock is aligned at right angles to the direction of motion of the space ship.

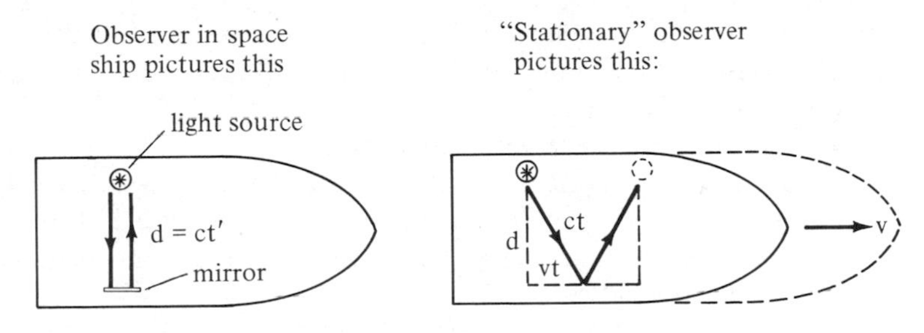

Figure 2.

The owner of the clock says the light flash requires time t' to reach the mirror and time t' to return, where

$$ct' = d.$$

But the stationary observer argues that, since the space ship was moving while the light flash was traveling to the mirror and back, the trip in each direction requires longer, say time $t > t'$. Specifically, the stationary observer argues that if the one-way trip for the light flash took time t, then the space ship moved forward a distance vt during that time. Thus, by the Pythagorean Theorem (see Figure 2),

$$(ct)^2 = d^2 + (vt)^2.$$

Substituting $d = ct'$ into this, one finds

$$c^2t^2 = c^2t'^2 + v^2t^2.$$

So $t'^2 = t^2 - (v^2/c^2)t^2$. Thus $t' = t\sqrt{1 - (v/c)^2}$, or

$$t = \frac{t'}{\sqrt{1 - (v/c)^2}}. \qquad (*)$$

(Note that, for this to make sense, one must always have $|v| < c$.)

Thus, while the space traveler's clock is registering a time interval t', the "stationary" clock advances by the larger amount t. Hence the stationary observer says that the space traveler's clock is "slow."

Example 1. If a space ship is traveling at 60% of the speed of light, how "slow" are its clocks?

Solution. Since $v = 0.6c$, a time period t for "stationary" observer will correspond to only

$$t' = t\sqrt{1 - (0.6)^2} = t\sqrt{0.64} = 0.8t.$$

So the "stationary" observer concludes that the space traveler's clocks are 20% slow.

Of course, the traveler in the space ship moving at constant velocity v has every right to consider *himself* or *herself* stationary and the *other observer* as moving at speed v (in the opposite direction). Then the traveler in the space ship will consider the other observer's clocks to be slow.

Without going into the justification, let it be said that physicists are generally convinced that other types of clocks will agree with the Einstein–Langevin clock. These include spring-driven clocks, weight and pendulum clocks, and even biological "clocks."

Thus in the previous example, if the space traveler's heart beats 60 times per minute (by his or her clock), the "stationary" observer will say that it is really 20% slow, and so is only beating 48 times per minute. So the "stationary" observer concludes that the space traveler is aging more slowly.

The experimental evidence which is usually cited to confirm that "moving" clocks really do run slower than "stationary" ones comes from the study of some elementary particles called mu-mesons or muons. Muons are unstable particles created by cosmic rays hitting the atmosphere. At rest muons have an average life of only 2.2 microseconds (2.2×10^{-6} seconds). In that time, even if they traveled at the speed of light, they could not go more than about

$$ct = 186{,}000 \times 2.2 \times 10^{-6} = 0.4 \text{ miles}$$

before decaying. But experimentally it is found that muons traveling near the speed of light go much farther. Hence they must live longer than 2.2 microseconds.

Example 2. If a muon is traveling at 99.9% of the speed of light, how long would it live and how far could it travel from the viewpoint of a stationary observer?

Solution. This is much like Example 1 except that the numbers are harder to handle (without a calculator). The following shows how to do it. Assume that in its own reference frame the muon has lifetime $t' = 2.2 \times 10^{-6}$ seconds. Then, since it is traveling at the speed $v = 0.999c$, its lifetime from the point of view of a stationary observer would be, by Eq. (∗),

$$\begin{aligned} t = \frac{t'}{\sqrt{1-(v/c)^2}} &= \frac{t'}{\sqrt{1-(0.999)^2}} = \frac{t'}{\sqrt{(1+0.999)(1-0.999)}} \\ &\simeq \frac{2.2 \times 10^{-6}}{\sqrt{2 \times 0.001}} = 2.2 \times 10^{-6} \times \sqrt{500} \\ &= 2.2 \times 10^{-6} \times \sqrt{5} \times 10 = 2.2 \times 2.24 \times 10^{-5} \\ &= 5 \times 10^{-5}, \end{aligned}$$

about 50 microseconds.

During this time the muon would travel about

$$186{,}000 \times 50 \times 10^{-6} = 9.3 \text{ miles}.$$

PROBLEMS

1. A spaceship leaves earth on a long journey at half the speed of light. By how much will the occupants of the space ship age in 10 years of time measured on earth?
2. The space ship in Example 1 passes us as its clocks read 12 noon. In our reference frame, how far away will it be when its clocks read 1 PM?
3. How fast must a space ship travel in order that its occupants will only age 10 years while 100 years passes on earth?
4. The following is quoted from a newspaper article of 1977 October 9:

   Stationary muons live only 2 microseconds, but the scientists made some muons circle a storage ring at 99.4% the speed of light, thus simulating a space journey, and what do you think happened? Lo and behold! the muons lived 64 microseconds, and this 32-fold increase in their lifespan is within 0.2% of the amount predicted by Einstein.

   Show that there must be a mistake in the article quoted above.
   (a) If the muons travel at 99.4% the speed of light, how long should they live?
   (b) If the muons live 64 microseconds, how fast must they have been traveling?

9.3. Length Contraction

After the downfall of universal time, you might ask whether length measurements are the same for different observers. The answer again will be no! A moving rod (pointed in the direction of motion) appears shorter than a "stationary" one.

Consider a traveler with an Einstein–Langevin clock in a space ship moving at constant speed v relative to a stationary observer. But this time let the clock be positioned lengthwise in the space ship so that the light flash travels to the mirror in the direction of motion of the ship, and the reflection returns in the opposite direction.

Let t be the time for the round trip of a light flash from the source to the mirror and back as seen by a stationary observer. For this round trip, according to Section 9.2, the stationary observer concludes that the clock in the space ship only registers the shorter time

$$t' = t\sqrt{1-(v/c)^2}.$$

One must allow for the possibility that the space traveler sees a different length for the clock, say d', from that "seen" by a stationary observer, say d. (Perhaps d' and d will turn out to be equal, perhaps not.)

Clock apparatus as seen by traveler with clock

d′

mirror

source

Clock apparatus as pictured by stationary observer

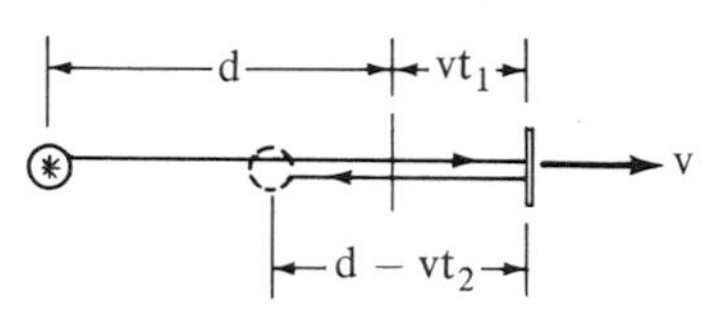

Figure 1.

Referring to Figure 1, the space traveler concludes that

$$t' = \frac{2d'}{c}.$$

But the stationary observer sees the ship moving with velocity v. So he or she concludes that while the light flash is traveling to the mirror, say in time t_1, the mirror moves away a distance vt_1. Thus $ct_1 = d + vt_1$. Similarly, while the reflected flash is returning to the source, say in time t_2, the source is moving toward it a distance vt_2. Thus $ct_2 = d - vt_2$. So the stationary observer concludes that the round trip of the light flash takes time

$$t = t_1 + t_2 = \frac{d}{c-v} + \frac{d}{c+v} = \frac{2cd}{c^2 - v^2}.$$

Now substitute this and $t' = 2d'/c$ into the equation $t' = t\sqrt{1-(v/c)^2}$ to find

$$\frac{2d'}{c} = \frac{2cd}{c^2 - v^2}\sqrt{1 - \left(\frac{v}{c}\right)^2}.$$

Thus

$$d\sqrt{1-\left(\frac{v}{c}\right)^2} = d'\frac{c^2 - v^2}{c^2} = d'\left[1 - \left(\frac{v}{c}\right)^2\right].$$

Finally divide both sides by $\sqrt{1-(v/c)^2}$ to get

$$d = d'\sqrt{1-(v/c)^2}. \qquad (*)$$

The stationary observer concludes that the moving rod is shorter than the length d' seen by the space traveler moving with the rod.

Now there is nothing special about the rod (which serves as part of the Einstein–Langevin clock) in this argument. So one must conclude that *all* moving objects are "contracted" (in the direction of motion) by the factor $\sqrt{1-(v/c)^2}$ from the point of view of a stationary observer. (Without proof, let it be noted that dimensions at right angles to the direction of motion are unchanged.)

The contraction of lengths in the direction of motion was actually proposed by G. F. Fitzgerald and later by H. A. Lorentz *before* Einstein's work. And a "principle of relativity" was anticipated by H. Poincaré in 1904.

The consequences of the Lorentz–Fitzgerald contraction for our space traveler are indicated in Figure 2. As far as the traveler is concerned, everything in the space ship is normal. But to a stationary observer, not only is the moving clock contracted, but the space ship itself and its occupants are flattened in the direction of motion by the factor $\sqrt{1-(v/c)^2}$.

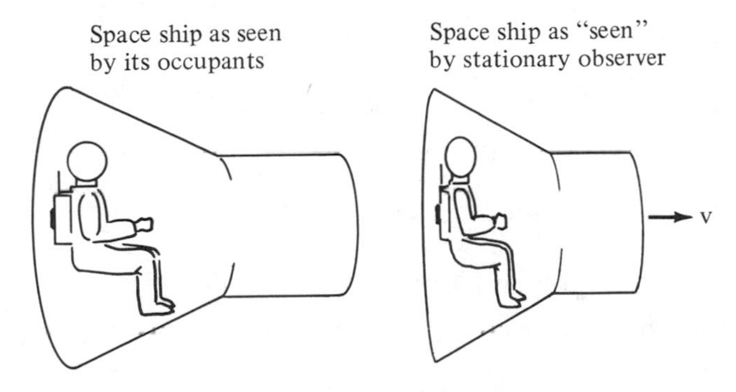

Figure 2.

Example 1. A space ship is 100 meters long, when measured at rest. How long would it appear to us (on earth) if it were passing by at 80% of the speed of light?

Solution. Given that $d' = 100$ meters and $v = 0.8c$, the length we would see is

$$d = d'\sqrt{1-\left(\frac{v}{c}\right)^2} = 100\sqrt{1-(0.8)^2} = 100\sqrt{0.36} = 100 \times 0.6 = 60 \text{ meters.}$$

Length contraction can now resolve some apparent "paradoxes" arising out of time dilation.

In Section 9.2 (Example 2) it was asserted that a muon traveling at speed v would live longer, from our point of view, than it would if at rest. Specifically, if its lifetime at rest were t', then we would see it actually living a time

$$t = \frac{t'}{\sqrt{1-(v/c)^2}}.$$

It has been determined that most muons produced in the upper atmosphere and moving toward earth actually reach the earth—traveling a distance of several miles. This is easily explained by an observer on earth who considers the muon to be living long enough to travel perhaps 9 miles. But to the muon, its lifetime is only about 2.2 microseconds. So how can the *muon* "explain" its ability to go any further than

$$ct = 1.86 \times 10^5 \times 2.2 \times 10^{-6} = 0.409 \text{ miles?}$$

Example 2. A typical muon produced by the action of cosmic rays at a height of 9 miles above the earth is moving toward the earth at speed $v = 0.999c$. If the muon's life, in its own reference frame, is only 2.2 microseconds, how do we explain its ability to reach earth, and how does the muon itself explain it?

Solution. From Example 2 of Section 9.2, the muon lives, in our reference frame,

$$t = \frac{2.2 \times 10^{-6}}{\sqrt{1 - (v/c)^2}} \cong 5 \times 10^{-5} \text{ seconds.}$$

Thus we are not surprised at its ability to travel a distance

$$vt \cong ct \cong 1.86 \times 10^5 \times 5 \times 10^{-5} = 9.3 \text{ miles.}$$

In its own reference frame, the muon is stationary and the earth is rushing toward it at speed $v = 0.999c$. Thus the muon concludes that the distance to earth (9 miles in our reference frame) is *contracted* to only

$$9\sqrt{1 - (v/c)^2} \cong 9\sqrt{20 \times 10^{-4}} = 9 \times \sqrt{20} \times 10^{-2} = 0.402 \text{ miles.}$$

It is not "surprised" to meet the earth in less than 2.2 microseconds, since in this time the earth would "travel"

$$v \times 2.2 \times 10^{-6} \cong 1.86 \times 10^5 \times 2.2 \times 10^{-6} = 0.409 \text{ miles.}$$

In closing this section, it should be remarked that the Lorentz–Fitzgerald contraction is considered to be a very real phenomena. But it is not something that one would literally see.

Consider a long truck driving down the road toward you, and you want to determine its length. If you look at the truck or take a snapshot of it, you do not see the contracted length. On the contrary, the truck will *appear* even longer than it would at rest. This is because your eye (or camera) notes the light arriving from the truck at just one instant. And the light you observe from the front of the truck, having a shorter distance to travel, reaches you more quickly than light from the back. Thus the image you see includes light emitted *fairly recently* from the front of the truck and light emitted *at an earlier instant* from the back. Thus the truck was further from you when the light left the back of the truck than it was when the light left its front. This makes the truck look longer.

Similarly, if the truck has passed you and is going away it will look short—for reasons which are quite distinct from the Lorentz–Fitzgerald contraction. In this case, the light from the front of the truck had to leave sooner (when the truck was closer to you) in order to reach your eye (or camera) at the same instant as light from the back.

If you really wanted to measure a truck moving at constant speed, you could do it with the help of a friend having a clock synchronized with your own clock. Station yourself and your friend alongside the road so that the front of the truck passes you when your clock reads 12 noon and the back of the truck passes your friend when his or her clock reads 12 noon. Then measure your separation from each other. This will be the length of the truck, and it will be less than the length if the truck were stationary. Specifically, you would find the length, as predicted by Eq. (*),

$$l\sqrt{1-(v/c)^2}$$

where l is the length of the truck at rest and v is its speed.

The presentation in Chapter 9 was adapted from T. M. Helliwell, *Introduction to Special Relativity* (Allyn and Bacon, Boston, Mass. 1966) and L. Marder, *Time and the Space Traveller* (George Allen & Unwin, London, 1971).

Problems

1. If a space ship which was known to be 70 meters long when at rest passes earth with a length of only 35 meters, how fast is it traveling?

2. Stanford University has a 2-mile-long linear accelerator (a "linac") for bringing elementary particles up to high speeds. Assume that an electron having passed through the first mile of the linac has reached speed $v = 0.9997c$, and assume this speed is constant for the remaining mile to the end of the linac.
   (a) How long will it take the electron to go the remaining mile in Stanford's reference frame?
   (b) How far does Stanford University have to go in the electron's reference frame?
   (c) How long does the rest of the trip take in the electron's reference frame?

3. A pi-meson (or pion) has a lifetime of about 2.5×10^{-8} seconds when at rest. If a pion is traveling at speed $v = 0.6c$,
   (a) how long will it live in our reference frame?
   (b) how far will it go in our reference frame?
   (c) how will the pion explain all this?

CHAPTER 10

Binary Arithmetic

Thousands of years before recorded history, herdskeepers kept track of their cattle or sheep by means of notches on a "tally stick." More efficient notations for the concepts "one," "two," "three," were developed by the ancient Egyptians, Romans, Chinese, and others. And these eventually evolved into the familiar decimal notation.

The various systems of notation used by ancient civilizations will not be discussed here. Rather the present chapter is primarily concerned with an alternative to the decimal system which has practical significance for the modern electronic computer and for data transmission links—the binary system.

10.1. Decimal, Binary, and Ternary Representation of Integers

The binary notation might best be made clear if it is introduced in comparison with the familiar decimal notation and one other similar possibility called ternary notation.

The decimal notation for positive integers uses ten familiar basic symbols—the digits 0, 1, 2, 3, 4, 5, 6, 7, 8, and 9. Integers greater than nine are represented by strings of these digits with the position of each digit determining whether it represents units, tens, hundreds, or thousands, etc. Thus the number written as 5047 represents

$$5 \text{ thousands} + 0 \text{ hundreds} + 4 \text{ tens} + 7 \text{ units}$$

or

$$5 \times 10^3 + 0 \times 10^2 + 4 \times 10 + 7 \times 1.$$

The use of a system so essentially based on ten and on powers of ten undoubtedly came about because humans have ten fingers.

Now consider a notation analogous to decimal notation, but using powers of two rather than powers of ten. Then the only digits are 0 and 1. Any integer is represented by a finite string of 0s and 1s. The digit on the far right represents the number of units—either 0 or 1. The digit next to it to the left indicates the number of twos. The next digit to the left gives the number of fours, and so on. Thus, for example, the number 1001 in binary notation represents

$$1 \text{ eight} + 0 \text{ fours} + 0 \text{ twos} + 1 \text{ unit}.$$

(Translated into decimal notation this is the number 9.) The number two itself is written as 10. Four is written as 100, and eight becomes 1000.

The decimal system is referred to as the base-ten system. Thus the binary system is the base-two system.

A base-three system or ternary system results if each positive integer is represented as a sum of units, threes, nines, twenty-sevens, etc. as follows. The digits used for this are 0, 1, and 2. A number larger than two is represented by a string of these digits, with the position of each digit in the array indicating what power of three it multiplies. For example, in the ternary system 1201 represents

$$1 \text{ twenty-seven} + 2 \text{ nines} + 0 \text{ threes} + 1 \text{ unit}.$$

(In decimal notation this number would be written as 46.)

The following table lists the first twelve positive integers in binary, ternary,

The First Few Positive Integers

Base Two				Base Three			Base Ten	
eights	fours	twos	units	nines	threes	units	tens	units
			1			1		1
		1	0			2		2
		1	1		1	0		3
	1	0	0		1	1		4
	1	0	1		1	2		5
	1	1	0		2	0		6
	1	1	1		2	1		7
1	0	0	0		2	2		8
1	0	0	1	1	0	0		9
1	0	1	0	1	0	1	1	0
1	0	1	1	1	0	2	1	1
1	1	0	0	1	1	0	1	2

and decimal form. In the space at the end of the list, you should extend each column to the next four integers. (Problem 2).

Even without the list, one can convert an integer from one form to another.

Example 1. Convert the decimal number 79 into (a) binary form and (b) ternary form.

Solution. (a) Rewrite the given number by successively extracting the highest power of two which it contains:

$$79 = 64 + 15 = 64 + 8 + 7 = 64 + 8 + 4 + 3 = 64 + 8 + 4 + 2 + 1.$$

Thus in binary notation the number is 1001111.

(b) Rewrite the given number by successively extracting the highest power of three or two times the highest power of three it contains:

$$79 = 2 \times 27 + 25 = 2 \times 27 + 2 \times 9 + 7 = 2 \times 27 + 2 \times 9 + 2 \times 3 + 1.$$

So in ternary notation the number is 2221.

The basic operations of arithmetic—addition and multiplication—are performed in base two or base three quite analogously to their mechanics in base ten. The computations are explained below primarily via examples.

Addition of base-two or base-three numerals is the same as in base ten, starting at the right-hand (units) column, except that "carrying" in each case means something different from its meaning in base ten.

For instance, if in a base-two addition problem the right-hand column adds up to two or three, the excess two is recorded by adding 1 to the twos column—the next column to the left. If the right-hand column adds up to four or five, the excess four is recorded by adding 1 to the fours column. Check the following examples to make sure you fully understand them.

Example 2. Addition computations in binary notation.

$$\begin{array}{r} 101 \\ 10 \\ 1100 \\ \hline 10011 \end{array} \qquad \begin{array}{r} 101 \\ 11 \\ 1101 \\ 110 \\ \hline 11011 \end{array}$$

When it comes to addition in base three, the procedure is again similar. But now one records an excess of three in the total of any column by carrying 1 to the next column to the left, and one records an excess of six by carrying 2 to the next column to the left. An excess of nine in any column total is recorded by carrying 1 *two* columns to the left, and so on.

Example 3. Addition in ternary notation.

$$\begin{array}{r} 12 \\ 2 \\ 110 \\ \hline 201 \end{array} \qquad \begin{array}{r} 222 \\ 122 \\ 22 \\ 121 \\ \hline 2111 \end{array}$$

For multiplication of numbers in binary or ternary notation, one again follows the same procedure as in base ten except for the meaning of "carrying." The following examples should explain the process.

Example 4. Multiplication in binary notation.

$$\begin{array}{r} 10110 \\ 1011 \\ \hline 10110 \\ 10110 \\ 10110 \\ \hline 11110010 \end{array}$$

Example 5. Multiplication in ternary notation.

$$\begin{array}{r} 211 \\ 102 \\ \hline 1122 \\ 211 \\ \hline 22222 \end{array}$$

PROBLEMS

1. Convert each of the decimal numbers 19 and 42 into equivalent (a) binary notation, and (b) ternary notation.
2. Extend the lists on page 157 to the next four integers.
3. Carry out the following additions in binary notation.
   (a) 1101 + 111 (b) 10111 + 111 + 101.
   (You can check your answers by converting to decimal form.)
4. Carry out the following additions in ternary notation.
   (a) 111 + 221 + 210 (b) 1220 + 222 + 212 + 220.
5. Perform the following multiplications in ternary notation.
   (a) 2111 × 21 (b) 212 × 21.
6. Perform the following multiplications in binary notation.
   (a) 1101 × 111 (b) 10111 × 111.

7. How many digits are required to express the number 10,000 (decimal) in (a) binary form? (b) ternary form?

8. Why do we prefer to use decimal notation rather than binary or ternary in everyday computations?

10.2. Subtraction and Division in Base Two

Now concentrate attention on the binary system.

After learning the notation for integers and the mechanics of addition and multiplication, the next topic is subtraction (of integers). Just as addition required attention to the new meaning of "carrying," so subtraction requires attention to the meaning of "borrowing." The following should illustrate the process.

Example 1. Subtraction in base-two notation.

$$\begin{array}{r} 100110 \\ 1001 \\ \hline 11101 \end{array}$$

The fourth operation of arithmetic is division, and this generally leads to fractions or other notations for noninteger numbers.

What do you suppose the symbol

$$0.1$$

means in binary notation?

Refer to the "point" in this symbol as the *binary point* (rather than the decimal point). The digit just to the right of the binary point represents the number of "halves" in the given number. The next digit to the right is the number of "fourths," and after that comes the number of "eighths."

Example 2. Convert to decimal equivalent the number given in base-two form as 101.1101.

Solution. In base ten this is

$$1 \times 2^2 + 0 \times 2 + 1 \times 1 + 1 \times \frac{1}{2} + 1 \times \frac{1}{2^2} + 0 \times \frac{1}{2^3} + 1 \times \frac{1}{2^4}$$
$$= 4 + 1 + \frac{1}{2} + \frac{1}{4} + \frac{1}{16} = 5\frac{13}{16} = 5.8125.$$

The meaning of binary notation can be described briefly as follows. Each "place" in base-two notation (to be filled by either 0 or 1) has a "value" which is half the "value" of the place immediately to its left; and the place just to the left of the binary point has the "value" 1.

Thus, multiplication by 10 (two) in binary notation is achieved by simply moving the binary point to the right one place. This moves each digit to a place with double the value and hence doubles the entire number.

For example, $0.1 \times 10 = 1.$ and $101.1101 \times 10 = 1011.101$. Similarly $0.01 \times 100 = 1$. Why?

Thus the following equivalences hold in base-two notation:

$$0.1 = \frac{1}{10},$$

$$0.01 = \frac{1}{100},$$

$$0.001 = \frac{1}{1000}.$$

And these look just like valid base-ten equations!

Now consider the problem of converting from decimal to binary notation.

Example 3. Convert the decimal number 17.72 to binary form both with and without the use of fractions.

Solution. Let us begin by successively subtracting off the highest power of 2 possible. Thus

$$\begin{aligned}
17.72 &= 16 + 1.72 \\
&= 2^4 + 1 + 0.72 \\
&= 2^4 + 1 + \frac{1}{2} + 0.22 \\
&= 2^4 + 1 + \frac{1}{2} + \frac{1}{8} + 0.095 \\
&= 2^4 + 1 + \frac{1}{2} + \frac{1}{2^3} + \frac{1}{16} + 0.0325 \\
&= 2^4 + 1 + \frac{1}{2} + \frac{1}{2^3} + \frac{1}{2^4} + \frac{1}{2^5} + 0.00125.
\end{aligned}$$

There is no reason to expect the binary form to terminate. If you drop the 0.00125, the binary form is 10001.10111 (rounded off).

Alternatively, one can find the exact equivalent of 17.72 in terms of base-two fractions. Since $17.72 = 17\frac{72}{100}$, it will suffice to convert the integers 17, 72, and 100 into their binary equivalents, and then reassemble the expression $17\frac{72}{100}$ using the binary symbols. Note that

$$17 = 16 + 1 = 2^4 + 1,$$

$$72 = 64 + 8 = 2^6 + 2^3,$$

and

$$100 = 64 + 32 + 4 = 2^6 + 2^5 + 2^2.$$

Thus, in binary notation these three integers are 10001, 1001000, and 1100100, respectively. Thus 17.72 is equivalent to the binary number

$$10001\frac{1001000}{1100100} = 10001\frac{10010}{11001}.$$

In the very last step the fraction was reduced by dividing numerator and denominator by the number 100 (four).

Now tackle division.

The process of division in base two is once again the analog of the same process in base ten. The next example demonstrates it.

Example 4. In base two compute $1001 \div 101$.

Solution

```
          1.110011
    101 ⟌1001
          101
          100 0
           10 1
            1 10
            1 01
               1000
                101
                 110
                 101
                   1
```

Thus the answer is a repeating binary: $1001/101 = 1.\overline{1100}$.

The maximum possible length of the repeating pattern could have been predicted. Since the divisor is 101 (five), there could be at most four different nonzero remainders. Thus the repeating pattern could be at most four digits long. (Compare with Problem 5 of Section 2.3.)

If you were not already acclimated to the decimal system, you would have to conclude that each step in the division process is *easier* in base two than it is in base ten. For instead of having to consider all the digits 0, 1, 2, ..., 9 as possible contributions to the quotient at each step, the only possibilities are 0 or 1.

Example 5 (for those who have studied Section 6.5). Check the result obtained in Example 4 by converting both the question and the answer to decimal form.

Solution. The question in base two ($1001 \div 101$) becomes $9 \div 5$ in decimal form. Moreover, $9 \div 5 = 1.8$.

The answer obtained in Example 4 in base two was $1.\overline{1100}$. This converts to the base-ten equivalent

$$1 + \frac{1}{2} + \frac{1}{2^2} + \frac{1}{2^5} + \frac{1}{2^6} + \frac{1}{2^9} + \frac{1}{2^{10}} + \cdots$$

$$= 1 + \left(\frac{1}{2} + \frac{1}{2^2}\right) + \left(\frac{1}{2} + \frac{1}{2^2}\right)\frac{1}{2^4} + \left(\frac{1}{2} + \frac{1}{2^2}\right)\frac{1}{2^8} + \cdots$$

$$= 1 + \frac{3}{4} + \frac{3}{4}\left(\frac{1}{16}\right) + \frac{3}{4}\left(\frac{1}{16}\right)^2 + \cdots.$$

This is 1 plus a geometric series which sums, as in Section 6.5, to

$$1 + \frac{3/4}{1 - 1/16} = 1 + \frac{12}{15} = 1 + \frac{4}{5} = 1.8.$$

Thus the result of Example 4 is confirmed.

Example 6. Convert the repeating binary number $x = 1.\overline{011}$ to an equivalent fraction using only base-two notation.

Solution. This can be handled in a manner analogous to that used in Section 2.3 for repeating decimals. Since the repeating pattern is three digits long, one should multiply x by 1000 and subtract x. Thus

$$\begin{aligned} 1000x &= 1011.011011\ldots \\ x &= 1.011011\ldots \\ \hline (1000 - 1)x &= 1010. \end{aligned}$$

Therefore $x = 1010/111$.

Problems

1. Perform the following subtractions in base two and check your answers by converting to base-ten notation.
   (a) $1010 - 101$ (b) $1001 - 111$.

2. Perform the following divisions in base two.
   (a) $1001 \div 11$ (b) $1 \div 11$ (c) $111 \div 101$.

3. Convert to decimal equivalents the following numbers given in binary form.
   (a) 1.01 (b) 10.011.

4. Convert the following base-ten numbers to equivalent binary form.
   (a) 5.125 (b) 5/6.

5. Convert the following repeating binaries to equivalent fractions *working entirely in base-two notation.*
   (a) $0.0\overline{1}$ (b) $1.0\overline{1100}$ (c) $0.\overline{1}$.
   (Do any of your final answers bear any relation to Problem 2?)

10.3. Applications

The obvious question remaining is: What is binary notation good for?

One important application arises in modern electronic computers and hand calculators. These devices are based on "switches" of one sort or another. And *the simplest type of switch is an on–off switch.*

The actual switches used in a computer are ultraminiature "solid-state" devices. But for simplicity, imagine a line of three ordinary on–off switches of any sort. Denote "on" for any switch by the digit "1," and denote "off" by the digit "0." Then any sequence of three digits, each being either 0 or 1, can be associated with one possible set of "off and on" positions for the three switches. For example, 110 would mean the switch on the left is "on," the one in the middle is "on," and the one on the right is "off," as in Figure 1.

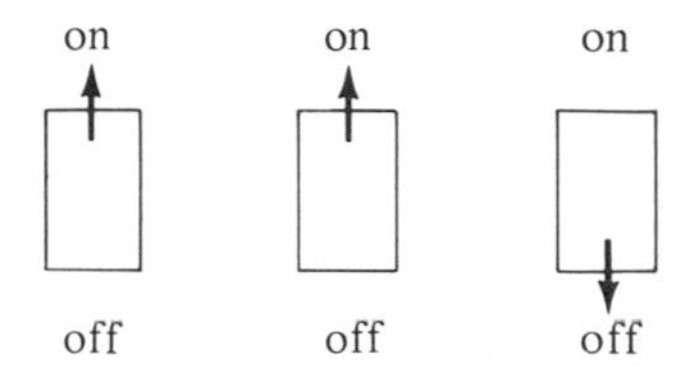

Figure 1.

So combinations of switch positions correspond to binary numbers. The above example (on, on, off) corresponds to 110 which is six. The line of three switches considered here can represent any integer from zero to seven (which isn't much).

To represent arbitrary integers just from zero to one hundred would require a group of seven on–off switches. Can you see why?

Thus a modern electronic computer contains the equivalent of millions of electronic switches. It does all its internal computations using binary arithmetic.

But, while the machine uses binary arithmetic internally, it is designed to receive human input in decimal form, which it then converts to binary. The computer finally converts the results of its computations back into decimal form for the convenience of the user.

A second, and related, important application of the binary system arises in the transmission of information.

A familiar tool for sending information over great distances is the radio. One can receive information in the form of high fidelity voice and music sounds over distances of 100 miles or more on an inexpensive radio receiver. Using the same basic principles, a TV set reproduces visual images from transmitters 30 or 50 miles away with good results. An important factor in this success is the power of the transmitter. The commercial radio and TV stations which broadcast the signals we receive at home are transmitting with powers up 50,000 watts.

If one tries to send radio messages with much weaker signals or over much greater distances, the problem becomes one of interference. The multitudes of man-made and nature-made electromagnetic signals which are all around us can mask the desired signal at the receiver. Unwanted signals are usually referred to as "static" or "noise."

This brings us back to the binary system.

Although you may not be able to understand a human voice on account of the static, it is comparatively easy to detect the difference between a signal being "on" and being "off." So, under difficult conditions one should think of transmitting messages by "on" and "off" signals.

Instead of using a simple "on and off" signal, the International Morse Code uses groups of "dots" and "dashes" to represent letters of the alphabet. An "amateur" radio operator with *as little as 5 watts* of transmitting power can sometimes communicate around the world using Morse code. This would not be possible in terms of voice reproduction.

Example 1. In order to be able to represent each letter of the English alphabet by appropriate positions of a group of on–off switches, how many switches would you need?

Solution. Since one must represent 26 different characters, there must be (at least) *five switches*. With five switches one can record binary numbers from 0 to 11111 (32 different combinations). Note that four switches would only provide 16 possible combinations.

As mentioned above, the Morse code uses groups of "dots" and "dashes" to represent letters. The first six letters of the alphabet are represented as follows:

A ·–
B –···
C –·–·
D –··
E ·
F ··–·

In fact, each of the 26 letters of the English alphabet is represented in Morse code by *at most four* dots and dashes! Why, in view of Example 1, does it not require five?

From the listing above for A through F, you see that not every letter uses four dots and dashes. The fact is that in Morse code one has available three basic "digits:" dot, dash, and off. So you might think the system is ternary. If it were, then to cover 26 distinct characters one would only need groups of three "digits." For this would cover the ternary numbers from 0 through 222, i.e., 27 different combinations. This would appear to be enough for our alphabet.

But is it?

Example 2. Show that one *cannot* represent all 26 letters of the English alphabet using sequences of three dots, dashes, and offs.

Solution. There is a practical problem. One cannot manage with only three digits because there is no way to signal "off" at the *beginning* of a character. More specifically, suppose we say "off" is 0, "dot" is 1, and "dash" is 2. And suppose everyone agreed that each character will be represented by a sequence of three digits. Then we could send the message 120 as dot–dash–off, and the receiver should recognize this as "A." But we have no way of signalling 012. The receiver would not know that anything was coming until he or she detected dot–dash. This would be indistinguishable from 120.

So one needs *more than three* digits to transmit the alphabet by Morse code.

Problem 3 pursues this question further to show that one indeed needs sequences of dots and dashes up to length four—but not five. (The full Morse code also has symbols for punctuation and numerals, and these do require strings of more than four dots and dashes.)

The end of a letter in Morse code is indicated, to the human receiver, by a pause. But in a modern electronic computer or communication system, this gimmick of pausing between characters would be inefficient.

Messages are transmitted within or between computers and over long-distance telemetry channels via binary signals, i.e., either "off" or "on." And some fixed basic time interval—perhaps 1/300 of a second—is assigned for each "off" and "on" signal, representing the digits 0 and 1, respectively. The receiving equipment also measures the time intervals, and if the signal is "on" for *twice* the basic time interval, then it counts as 11 ("on–on"). If it remains "off" for twice the basic time interval, it is interpreted as 00.

In binary usage, each 0 or 1 (each "off" or "on" in electrical terms) is referred to as one **bit** of information. And, in a modern digital computer or communications system, a fixed number of bits—perhaps eight—is assigned to each "character" or **byte**. Then the machine simply runs all the characters together in a string, with each bit taking the allotted time interval. The receiver, another computer connected to the first either by wire or radio transmission, must measure time intervals in order to know when one bit ends and another begins.

Example 3. In the IBM System/370 every character is represented by eight bits. The signals for a space and for the first six capital letters of the alphabet are

space	01000000
A	11000001
B	11000010
C	11000011
D	11000100
E	11000101
F	11000110

Interpret the received message

1100001011000001110001000100000011000100110001011100010111000100.

Solution. Mark off each block of eight bits. Then note that the first eight bits, 11000010, represent B, and the next eight bits, 11000001, represent A. Continue in this manner to find that the complete message is BAD DEED.

The system mentioned in Example 3 uses eight bits per character, instead of the five as suggested by Example 1. The reason for this is that one wants to be able to represent capitals *and* lowercase letters, decimal digits (0, 1, 2, ..., 9), punctuation marks, *and* various mathematical symbols, *and* instructions for the computer itself. If each character contains eight bits, how many distinct characters can be represented? (Problem 1.)

As noted earlier, binary signals are used in long-distance communication in order to eliminate the problem of interference ("static").

But even data in binary form can acquire errors during transmission. So means have been developed for detecting (and even correcting) such errors. The simplest idea of this type is as follows.

Example 4. Suppose each character in a certain coding system is represented by eight bits. Let us add one more bit at the end of each character (making a total of nine bits), with this extra bit being 0 if the character contains an odd number of 1s, and being 1 if the character contains an even number of 1s. This additional bit is called a **parity bit**.

Thus, for example, an original character 10011010 would be transmitted as 100110101, and the character 00101010 would be transmitted as 001010100.

If you ever received a character 100110100, you would know it was *false*. Why?

On the other hand, if you received 100110101, it *probably* truly represents 10011010. More precisely, the received signal *cannot* contain an error in just *one* bit. And in a good communication link it is very unlikely that a single character would simultaneously acquire errors in more than one bit.

In practice, much more sophisticated means are used for treating errors. By the use of "Hamming codes" errors can not only be detected, but also corrected, by electronic computers.

Some of the more spectacular examples of long-distance transmission of data occur in the space program. For example, in October and November 1980, Voyager 1 sent back to earth beautiful pictures of Saturn and its rings and moons, plus multitudes of other data. The distance was about *one billion miles*, and the transmitter on Voyager 1 had an output of *only 20 watts*.

This was possible because *all* the information, including the pictures, was coded into binary form—i.e., on–off signals—for thousands of dots. The transmitter on Voyager 1 sent back 45,000 bits per second to be processed by computers on earth into pictures and other forms usable by humans. This processing takes time. So, much of the pictorial and other information sent

Courtesy of NASA

Saturn as seen by Voyager 1 on October 18, 1980.

back had to wait (recorded magnetically on tape as billions of 0s and 1s) for as long as 3 months until its turn came for time on the computer for processing.

It should be conceded that the processing which produced the spectacular pictures involved some "computer enhancement" beyond the basic reconstruction of the dots.

Problems

1. How many distinct characters can be represented if one always uses eight bits per character? (Compare Examples 1 and 3.)
2. We all know how to count to ten on our fingers. Explain how just ten fingers, used to represent binary digits, could serve to count to more than one thousand.
3. How many different characters could you represent *using dots and dashes* (short and long tones) if each character consisted of
   (a) just one tone?
   (b) one tone or a sequence of two tones?
   (c) one tone or a sequence of two or three tones?
   (d) one tone or a sequence of two, three, or four tones?
   (e) one tone or a sequence of two, three, four, or five tones?
   Does this show why the Morse code can represent each letter of the English alphabet using sequences of no more than four dots and dashes?

4. Assuming that eight-bit characters are augmented by a ninth parity bit at the end, as in Example 4, which of the following received signals is definitely false?
   (a) 111011011 (b) 101001111 (c) 101010010 (d) 111010010

5. Assume that each character in the coding system of Example 3 is augmented by a parity bit at the end, as in Example 4. Interpret each of the following received English words (assuming that only the letters listed in Example 3 are involved).
   (a) 110001101110000010110001000110001011,
   (b) 110001101110001011110000011110001000,
   (c) 110001100110001011110000010110001000.
   [*Hint*. First mark off blocks of nine bits each, then check for errors, then translate.]

CHAPTER 11

Sets and Counting

This brief chapter provides the prerequisites for Chapter 12 on probability and Chapter 13 on cardinality. For these topics one needs some acquaintance with the mathematical notation for sets.

Furthermore, in elementary probability examples it is often important to determine the number of different ways a group of objects can be selected and/or ordered.

11.1. Set Notation

A **set** is any well-defined collection of objects or concepts, called the **members** (or **elements** or **points**) of the set. A set S can be described by listing its members in braces $\{\ \}$. For example, $\{1, 2, 3, 4, 5, 6\}$ represents the set containing the first six positive integers, {a, e, i, o, u} is the set of vowels in the English alphabet, and {h, t} is the set of possible outcomes when one tosses a coin using h for "heads" and t for "tails."

The order in which the members of a set are listed is immaterial. Thus the first set above is the same as $\{2, 1, 4, 5, 3, 6\}$ and the third is equivalent to {t, h}.

Often a set contains many members, and one prefers not to take the time or space to list them all. Then, if there is no danger of confusion, one can omit some members of the list and denote the omission by three dots. Thus

$$\{1, 2, 3, \ldots, 100\}$$

represents the set of all integers from 1 to 100, inclusive. And the expression

$$\{a, b, c, \ldots, z\}$$

might be used to represent the set of all letters of the English alphabet.

Another useful format for describing sets is

$$\{x\text{: property or properties of } x\}.$$

Here x (or other convenient symbol) represents a member of the set. Then comes a colon followed by the exact property or properties which characterize a member of the set.

For example,

$$\{n\text{: } n \text{ is an integer, } 1 \le n \le 100\}$$

completely describes the set of all integers from 1 to 100, inclusive; and

$$\{x\text{: } x^2 = 2\}$$

describes the set $\{\sqrt{2}, -\sqrt{2}\}$.

Example 1. Describe the set $\{y\text{: } y \text{ is an integer, } 0 \le y \le 4\}$ in two other different but equivalent forms.

Solution. Among many possible equivalent descriptions of this set, two are

$$\{0, 1, 2, 3, 4\} \qquad \text{and} \qquad \{x\text{: } x \text{ an integer, } 0 \le x < 5\}.$$

A set may contain just one member. For example $\{5\}$ is the set containing just the number 5. And it is sometimes important to make a distinction between the *number* 5 and the *set* $\{5\}$.

Example 2. Describe the set of all possible outcomes when one simultaneously tosses a penny and a dime.

Solution. Writing h for heads and t for tails, the possible outcomes can be described as ordered pairs of symbols h and/or t. Let the first symbol in an ordered pair refer to the penny and the second to the dime. Thus (h, t) means the penny came up heads and the dime came up tails. The set of all possible outcomes is then

$$\{(\mathrm{h}, \mathrm{h}), (\mathrm{h}, \mathrm{t}), (\mathrm{t}, \mathrm{h}), (\mathrm{t}, \mathrm{t})\}.$$

Example 3. Describe the set of all possible outcomes when one simultaneously tosses two "identical" pennies.

Solution. You might be tempted to say that there are now only three possible outcomes—two heads, two tails, or one of each. But for the forthcoming probability applications it is essential to say that there are four possible outcomes, just as in Example 2.

Note that, no matter how identical the two pennies look, they still have their own "identities," and they cannot switch identities. So, if penny No. 1 is heads and penny No. 2 is tails, that should be considered different from the case when No. 1 is tails and No. 2 is heads.

Two "identical" coins must be considered different.

There are two standard symbols used for combining sets.

The symbol $\cup$, pronounced, **union**, linking two or more sets means form the new set composed of all the elements which belong to either, or any, of the given sets. So, if A and B are two sets, then $A \cup B$ represents the new set composed of everything in A together with everything in B.

Example 4. Let $A = \{1, 2, 3\}$, $B = \{2, 3, 5\}$, and $C = \{1, 7\}$. Then

$$A \cup B = \{1, 2, 3, 5\},$$

$$A \cup C = \{1, 2, 3, 7\},$$

$$B \cup C = \{1, 2, 3, 5, 7\},$$

and

$$A \cup B \cup C = \{1, 2, 3, 5, 7\}.$$

The symbol $\cap$, pronounced **intersect** (or **intersection**), linking two or more sets means form the new set composed of just those elements which belong to both, or all, of the given sets. Thus, if A and B are two sets, then $A \cap B$ is the new set composed of those elements which are common to both A and B.

Example 5. For the sets A, B, and C of Example 4

$$A \cap B = \{2, 3\} \quad \text{and} \quad A \cap C = \{1\}.$$

A special set is

$$\emptyset, \text{ the } \textbf{empty set},$$

which contains no elements.

Example 6. For the sets of Example 4,

$$B \cap C = \emptyset \quad \text{and} \quad A \cap B \cap C = \emptyset.$$

Two sets are said to be **disjoint** if they have no points in common. Since $B \cap C = \emptyset$ in Example 6, B and C are disjoint.

To denote that x is a **member of** a set A, one writes $x \in A$. And $x \notin B$ means that x is *not* a member of B. For the sets of Example 4, $1 \in A$ and $1 \notin B$.

Set A is said to be a **subset** of set B, written $A \subset B$, if every member of A also belongs to B. Again referring to the sets of Example 4, $A \subset B \cup C$.

For any set A, one says

$$A \subset A \qquad \text{and} \qquad \emptyset \subset A.$$

In each particular example the sets considered will all be subsets of some "universal set" or "space" S. Then, for any given set A, the **complement of** A, denoted by A^c is the set of all members of S which are not in A. In symbols,

$$A^c \equiv \{x \in S: x \notin A\}.$$

So, for any set A,

$$A \cap A^c = \emptyset \qquad \text{and} \qquad A \cup A^c = S.$$

For example, if the sets in Example 4 are regarded as subsets of the space $S = \{1, 2, 3, 4, 5, 6, 7, 8, 9, 10\}$, then

$$B^c = \{1, 4, 6, 7, 8, 9, 10].$$

The specific examples used thus far have all involved sets with just a finite number of elements (or members). Some examples of sets containing infinitely many members are

$\{1, 2, 3, \ldots\}$, representing all the positive integers,
$\{x: 3 \leq x \leq 5\}$, the set of all numbers from 3 to 5 inclusive, and
$\{r: r \text{ rational}, 0 \leq r < 1\}$, the set of all rational numbers from 0 to 1, including 0 but excluding 1.

Example 7. Let $A = \{x: x \leq 1\}$ and $B = \{x: x \geq 0\}$ be sets in the space S of all real numbers. Then

$$A \cup B = S, \qquad A \cap B = \{x: 0 \leq x \leq 1\}, \qquad \text{and} \qquad A^c = \{x: x > 1\}.$$

PROBLEMS

1. Rewrite each of the following sets in two other equivalent forms: $A = \{2, 4, 6\}$, $B = \{x: (x-2)(x-1) = 0\}$, $C = \{2, 3, 5, 7, 11\}$.
2. Consider $A = \{(\text{h}, \text{h}), (\text{t}, \text{h})\}$ and $B = \{(\text{h}, \text{t}), (\text{t}, \text{h}), (\text{t}, \text{t})\}$ to be subsets of the space $S = \{(\text{h}, \text{h}), (\text{h}, \text{t}), (\text{t}, \text{h}), (\text{t}, \text{t})\}$ (motivated by Examples 2 and 3). Find $A \cup B$, $A \cap B$, A^c, and B^c.
3. Describe the set (or space) S of all possible outcomes when one tosses a coin three times.
4. Describe three subsets of the set S of Problem 3, such that each pair of these three sets is disjoint.
5. Simplify $\{x: x \leq 0\} \cap \{x: (x-3)(x+2) = 0\}$.

11.2. Counting

Often one is more interested in knowing the number of elements in a set than in seeing a listing of those elements. This section introduces techniques for determining the number of elements in a set without necessarily writing them all down.

For a set A with finitely many members, $n(A)$ will denote the number of members of A.

Note that the entire symbol $n(A)$ represents a number. The parentheses do *not* signify multiplication in this usage. (An alternative notation might have been n_A; but $n(A)$ is more traditional.)

Example 1. How many members are in the set of all possible outcomes when one rolls a pair of dice—one white (with black spots) and the other black (with white spots).

Solution. Represent each possible outcome by an ordered pair of integers from 1 to 6 inclusive. Let the ordered pair (i,j) represent the outcome in which i shows on the white die and j on the black die. Then the set of all possible outcomes can be represented as

$$\begin{aligned} S = \{&(1,1), (1,2), (1,3), (1,4), (1,5), (1,6),\\ &(2,1), (2,2), \quad . \quad . \quad . \quad (2,6),\\ &\;\vdots \qquad\qquad\qquad\qquad\qquad\quad \vdots\\ &(6,1), (6,2), \quad . \quad . \quad . \quad (6,6)\}. \end{aligned}$$

How could you determine the number of elements in this set *without* writing them down?

Reason as follows. Each possible outcome is described when two "slots"

$$\_\_ \quad \_\_$$

are filled, in order. The first is to be filled by the number 1, 2, 3, 4, 5, or 6, representing the white die, and the second is to be filled by an integer from 1 to 6 for the black die. For each of the six possible ways of filling the first slot, there are six possibilities for the second. This means a total of $6 \cdot 6 = 36$ pairs. So $n(S) = 36$.

Example 2. How many members are in the set of all possible outcomes when one rolls a pair of "identical" white dice?

Solution. No matter how identical the dice may appear, they still have their own identities (as did the two "identical" pennies earlier). Thus, for example, 5 on die No. 1 and 3 on die No. 2 must be considered different from 3 on No. 1 and 5 on No. 2.

Again the answer is 36, just as in Example 1.

Example 3. If one tosses three "identical" pennies, how many outcomes are possible?

Solution. Once again the "identical" pennies should be considered as somehow distinguishable. Call them penny No. 1, penny No. 2, and penny No. 3.

Each possible outcome can now be represented by an ordered triple of h's and t's. The first entry in such a triple will be the showing on penny No. 1, the second the showing on penny No. 2, and the third the showing on penny No. 3. Thus, for example, (h, t, t) means heads on No. 1, tails on No. 2, and tails on No. 3.

To determine how many possible triples there are, think of filling three slots in order ___ ___ ___. The first must be filled by either h or t, two possibilities. For each of these choices, there are two possibilities for the second slot, making a total of $2 \cdot 2 = 4$ cases. Then for each of these four cases, there are two possibilities for filling the third slot—a grand total of $2 \cdot 2 \cdot 2 = 8$ possible outcomes.

Example 4. In how many different ways can one order (or permute or rearrange) the five letters a, b, c, d, e?

Solution. If S is the set of all possible orderings or "permutations" of these five characters, then each member of S is an ordered quintuplet. Examples are (a, b, c, d, e), (b, d, c, a, e), and (d, a, b, e, c).

To determine how many such elements there are, note that each is obtained by filling five slots in order, ___ ___ ___ ___ ___. But now, in contrast to Examples 1, 2, and 3, each available character, a, b, c, d, and e, can be used only once.

You can choose any one of the five characters to fill the first slot. But, whenever a choice is made for the first slot, only four options remain for the second slot. So, in filling the first two slots, there are $5 \cdot 4$ possible outcomes.

Now, regardless of the characters used in the first and second slots, there remain only three to choose from for the third slot. This gives a total of $5 \cdot 4 \cdot 3$ permutations considering just the first three slots.

Continuing with this reasoning, there must be $5 \cdot 4 \cdot 3 \cdot 2 \cdot 1$ possible permutations altogether. Thus the number of elements in S is

$$n(S) = 5 \cdot 4 \cdot 3 \cdot 2 \cdot 1 = 120.$$

In general, the number of permutations (or orderings) of k distinct objects is always $1 \cdot 2 \cdot 3 \cdots k$. This product is denoted by $k!$, pronounced "k factorial."

So the answer for Example 4 could be written, $n(S) = 5!$.

Example 5. A poker hand is a set of five playing cards from a standard deck of 52 distinct cards. How many different poker hands are possible?

Solution. The content and value of a poker hand has nothing to do with the way the cards are ordered. However, it is easiest to first answer the question, how many different *ordered* poker hands are possible?

This is answered via reasoning analogous to that in Example 4. There are five slots to fill, and for the first slot 52 different choices are possible. But having made this choice, only 51 cards remain from which to fill the second slot. So the first two slots can be filled (in order) in

$$52 \cdot 51 \qquad \text{different ways.}$$

Now, regardless of how the first two slots were filled, there remain 50 cards from which to choose the filler of the third slot. Thus, the first three slots can be filled in

$$52 \cdot 51 \cdot 50 \qquad \text{different ways.}$$

Continuing thus, one concludes that there are

$$52 \cdot 51 \cdot 50 \cdot 49 \cdot 48 \qquad \text{different ways}$$

of forming *ordered* poker hands.

But, from the point of view of poker, different orderings of a given group of five cards are of no interest. So there are nowhere near as many different "hands" as there are different "ordered hands." In fact, Example 4 shows that any group of five distinct objects can be ordered or permuted in 5! different ways. So for each poker hand we must have actually counted 5! ordered hands. If n is the number of possible poker hands, we have computed $n \cdot 5!$.

Thus, the number of different poker hands (without regard to order) is

$$n = \frac{52 \cdot 51 \cdot 50 \cdot 49 \cdot 48}{5!} = 2{,}598{,}960.$$

More generally, if one wants to select a **permutation**, i.e., an ordered list, of k items from a group of p distinct objects, this can be done in

$$p(p-1)(p-2)\cdots(p-k+1) \qquad \text{different ways.}$$

If one merely wants any **combination** (without regard to order) of k items from a group of p distinct objects, this can be chosen in

$$\frac{p(p-1)(p-2)\cdots(p-k+1)}{k!} \qquad \text{different ways.}$$

Problems

1. How many members are in the set of possible outcomes when one tosses a coin four times?
2. Describe the set of possible outcomes when one tosses a coin and rolls a die. How many members does this set have?

3. How many possible outcomes are there if one rolls three dice? (As in previous examples, you should imagine the dice to be individually identifiable.)

4. (a) In how many ways can one draw 3 cards—a first, second, and third in order—from a standard deck of 52 distinct cards? (The cards drawn are not replaced.)
(b) Among all the *ordered* drawings of three cards in part (a), how many would result in three aces?

5. (a) In how many ways can one deal a combination of 3 cards (without regard to order) from a standard deck of 52 distinct cards?
(b) How many of these "hands" of three cards would be composed of three aces?

6. Show that the number of "ordered poker hands" found in Example 5 can be written as $52!/47!$, and the number of different "hands" without regard to order is $52!/(47!\,5!)$.

CHAPTER 12

Probability

Almost everyone uses the language and ideas of probability:

"There is a 20% chance of rain tomorrow."

"The probability that the baby will be a boy is slightly more than half."

"What is the probability of passing this course?"

The following pages will explain some of the language and results of probability theory, and illustrate its uses in quality control, life insurance, smoking-vs.-health statistics, and of course gambling.

12.1. Elementary Ideas and Examples

What do we mean when we say that a tossed coin is just as likely to come up "heads" as "tails?"

It should not be expected that tossing a coin twice will produce one head and one tail, nor that 10 tosses will necessarily yield five heads and five tails.

However, if the coin is a "fair coin"—symmetric and balanced—then it ought to be true that in a *large number of tosses*, say 100, heads will appear *approximately* half the time. Indeed, you could convince yourself that this is the case by experimenting for a few minutes. One then says, "the probability of getting heads is $\frac{1}{2}$."

Similarly, when a fair die is rolled, each of the six possible outcomes (1, 2, 3, 4, 5, and 6) is "equally likely." This means that if it is rolled *many* times it should produce each of these six numbers on approximately one-sixth of the rolls. So, for example, "the probability of rolling 1 is $\frac{1}{6}$."

When a coin is tossed, one refers to "heads" and "tails" as the two "basic possible outcomes" of the experiment. When a die is rolled, the basic possible outcomes are 1, 2, 3, 4, 5, and 6.

The set of all basic possible outcomes for an experiment or a natural occurrence is called the **sample space** S.

To each basic possible outcome, one would like to assign a number called the probability of that outcome.

If, as in the case of a fair coin or a fair die, each basic possible outcome is "equally likely," the **probability** $P(a)$ assigned to a basic possible outcome a is defined to be

$$P(a) = \frac{1}{\text{number of equally likely basic possible outcomes}} = \frac{1}{n(S)}.$$

You might find this an unsatisfactory "definition" of $P(a)$ since the expression "equally likely" was not defined. Further comments on this follow Example 2 below.

Example 1. Let h for heads or t for tails represent the side which comes up when a fair coin is tossed. Then the sample space is {h, t} where h and t are equally likely. So one concludes that the probability of getting heads is

$$P(\text{h}) = \tfrac{1}{2}.$$

Similarly,

$$P(\text{t}) = \tfrac{1}{2}.$$

Note that $P(\text{h})$ and $P(\text{t})$ are just convenient notations for certain numbers (in this case $\frac{1}{2}$). The parentheses in these symbols do *not* imply multiplication. In what follows, you will encounter other expressions such as $P(2)$, $P(x > 4)$, and $P(3 \text{ of diamonds})$. In each case, the entire symbol just represents a number.

Example 2. When a fair die is rolled $S = \{1, 2, 3, 4, 5, 6\}$ and $P(1) = P(2) = P(3) = P(4) = P(5) = P(6) = \frac{1}{6}$.

The assigned probabilities above are intuitive concepts. They are interpreted to mean that if a coin is tossed *many times*, then *approximately* half of these tosses will produce heads; and if a die is rolled *many times* (perhaps 1000 times), then *approximately* one-sixth of these rolls will yield a 1.

The intuitively assigned probabilities are useful only if experiments confirm these interpretations.

Example 3. Shuffle and cut a standard deck of cards. Then, since the deck contains 52 distinct cards and each is equally likely to end up on top, the probability that the top card will be the three of diamonds ought to be

$$P(3 \text{ of diamonds}) = 1/52.$$

Again this means that if the experiment (shuffling, cutting, and noting the top card) were repeated, say, 1000 times, then on approximately 1000/52 occasions the top card should be the three of diamonds.

Example 4. Let x be the number which comes up when a fair die is rolled, and consider the probability of getting 5 or 6. This can be written as $P(x = 5 \text{ or } 6)$ or $P(x > 4)$. Thinking of the many repeated (imaginary) experiments, it appears natural that

$$P(x > 4) = \tfrac{1}{6} + \tfrac{1}{6} = \tfrac{1}{3},$$

since exactly two of the equally likely basic possible outcomes are acceptable. (This means that in a large number of rolls of the die, x should be greater than 4 *about* one-third of the times.)

Similarly, the probability that x is odd can be written $P(x \text{ is odd})$ or $P(x = 1, 3, \text{ or } 5)$; and, one reasons,

$$P(x \text{ is odd}) = \tfrac{1}{6} + \tfrac{1}{6} + \tfrac{1}{6} = \tfrac{1}{2}.$$

This implies the expectation that if a fair die is rolled many times, about half of those rolls will yield an odd number.

The experiments considered thus far—tossing a coin, rolling a die, and drawing a card—are just about the simplest possible experiments, and the probabilities described should not be surprising.

Experiments or natural occurrences having more complicated possible outcomes can be less obvious and hence more interesting.

Example 5. If two coins are tossed, what is the probability of getting two heads?

Solution. Here one must be careful. If the "equally likely" basic possible outcomes were two heads, two tails, and one of each, then the answer would be $\frac{1}{3}$. But this is *not* the case.

As in Chapter 11, you must consider the coins to be individually identifiable. Call them coin No. 1 and coin No. 2. Then the sample space becomes

$$S = \{(\text{h}, \text{h}), (\text{h}, \text{t}), (\text{t}, \text{h}), (\text{t}, \text{t})\},$$

where the first position in each ordered pair is assigned to coin No. 1 and the second position is assigned to coin No. 2. And the four basic possible outcomes in S are equally likely. So

$$P(\text{h}, \text{h}) = \tfrac{1}{4}.$$

Henceforth, the basic possible outcomes when two coins are tossed will be denoted more briefly as hh, ht, th, and tt. So the sample space will be written, $S = \{\text{hh}, \text{ht}, \text{th}, \text{tt}\}$.

A basic possible outcome or a set of basic possible outcomes for an experiment or a natural occurrence will be called an **event**. Thus, for any particular experiment, every event is a subset of the sample space.

In particular, the sample space itself is an example of an event.

In the experiment of tossing two coins, one can describe many possible events. For example, the following sets are events:

{hh}	(containing a single basic possible outcome),
{ht}	(containing a single basic possible outcome),
{ht, th}	(containing two basic possible outcomes),
{ht, th, tt}	(containing three basic possible outcomes).

Example 6. If two coins are tossed, what is the probability of getting one head and one tail without regard to which comes first?

Solution. Let x denote the actual outcome from among the four members of the sample space. Then one head and one tail means either $x = \text{ht}$ or $x = \text{th}$. Define the event

$$A = \{\text{ht}, \text{th}\}.$$

So one head and one tail means $x \in A$, and the desired probability is

$$P(x \in A).$$

Since ht and th are two of the four equally likely basic outcomes, one would expect the actual outcome x to be a member of A in approximately half of the two-coin experiments. Thus one defines

$$P(x \in A) = \tfrac{1}{2}.$$

For even greater brevity, this is commonly written as

$$P(A) = \tfrac{1}{2}.$$

In general, if A is an event (a set of basic possible outcomes), one would like to define the "probability" that $x \in A$, denoted by $P(A)$, to have the following property. If the experiment is repeated m times where m is a large number, then the number of occasions in which the outcome is a member of A is approximately

$$m \cdot P(A).$$

This generalizes the notion of probability of a single basic possible outcome.

Since $0 \leq m \cdot P(A) \leq m$, $P(A)$ must always satisfy

$$0 \leq P(A) \leq 1. \tag{1}$$

If S is the sample space of some experiment, then every repetition of the experiment will yield a point in the event S. So one *must* always define

$$P(S) = 1. \tag{2}$$

At the other extreme, if the event is $\emptyset$, the empty set, then no basic possible outcome will ever belong to $\emptyset$. So one *must* have

$$P(\emptyset) = 0. \tag{3}$$

In many elementary examples, one begins with a known sample space S containing $n(S)$ equally likely basic possible outcomes. Let A be a subset of S containing $n(A)$ basic possible outcomes. Then if the experiment is repeated m times (many times), one would expect to get an outcome in A approximately $m \cdot n(A)/n(S)$ times. This motivates the "definition"

$$P(A) = \frac{n(A)}{n(S)}.$$

Notice how easily this would give the result of Example 6.

Example 7. If a couple plans to have three children, what is the probability that all three will be the same sex? Assume here and in the future that the probability of having a girl is the same as the probability of having a boy (which is not quite correct).

Solution. Each possible outcome can be described by an ordered triple of b's and g's, in order of birth. The sample space is

$$S = \{\text{bbb, bbg, bgb, bgg, gbb, gbg, ggb, ggg}\},$$

and each of the eight basic possible outcomes will occur with probability $\frac{1}{8}$. The problem asks for $P(A)$ where $A = \{\text{bbb, ggg}\}$. And the answer is

$$P(A) = \frac{n(A)}{n(S)} = \frac{2}{8} = \frac{1}{4}.$$

Problems

1. If a card is drawn from a standard deck, what is the probability that it is (a) a jack, (b) a heart, (c) a jack or a heart?
2. Describe three more events (other than those on page 181) for the experiment of tossing two coins.
3. If three coins are tossed, (a) how many points are in the sample space of basic possible outcomes? (b) Describe the event (or set) A which corresponds to getting two heads and one tail. (c) What is the probability of getting two heads and one tail?
4. If a couple plans to have three children, what is the probability of their having two girls and a boy?
5. If a family contains four children, which is more probable: all the same sex, two of each sex, or three of one sex and one of the other?
6. If a coin is tossed and a card is drawn (from a well-shuffled standard deck), (a) how many points are in the sample space? (b) what is the probability of getting tails and the seven of clubs? (c) What is the probability of getting tails and a seven?
7. Your office is on the ninth floor of a 12-story building, but it seems that whenever you want to go down, the next elevator is going *up*. Is there anything to this?

12.2 Mutually Exclusive Events

If two events A and B, for some experiment, are disjoint, they are said to be **mutually exclusive**. (If an outcome belongs to A it cannot also belong to B.)

The simplest cases are those involving a clearly defined sample space S containing $n(S)$ equally likely members. Then, if A and B are two mutually exclusive events in S,

$$n(A \cup B) = n(A) + n(B).$$

Dividing each term of this equation by $n(S)$ leads to the conclusion

$$P(A \cup B) = P(A) + P(B).$$

It would be nice if this "additive property" always held for mutually exclusive events A and B. But it cannot always be justified by the simple argument given above because there are cases in which the basic possible outcomes of an experiment are *not* equally likely. Consider, for example, a die which is unbalanced so that the probability of getting 6 is greater than the probabilities of getting 1, 2, 3, 4, or 5. (This would be called an unfair or biased die.) In this case, one could not decide the value of $P(6)$ on the basis of intuition. Instead, one would have to resort to experiment—repeated many times—to suggest values of $P(1)$, $P(2)$, ..., and $P(6)$.

So, in order to assure that $P(A \cup B) = P(A) + P(B)$ whenever A and B are mutually exclusive events, one simply *postulates* it. What this does is impose a restriction on the possible choices for the probabilities in any particular situation. An expanded form of this postulate is stated below.

Events $A_1, A_2, \ldots,$ and A_k in some sample space, S, are said to be **mutually exclusive** if

$$A_i \cap A_j = \emptyset \qquad \text{whenever } i \neq j.$$

Postulate. *If $A_1, A_2, \ldots,$ and A_k are mutually exclusive events,*

$$P(A_1 \cup A_2 \cup \cdots \cup A_k) = P(A_1) + P(A_2) + \cdots + P(A_k). \tag{4}$$

This postulate, to be called Rule (4), is added to Rules (1), (2), and (3) of Section 12.1.

In particular, if S is a sample space of equally likely basic possible outcomes, then each of these outcomes will occur with probability $1/n(S)$. So for an event A containing k basic possible outcomes, say $a_1, a_2, \ldots,$ and a_k,

$$P(A) = P(a_1) + P(a_2) + \cdots + P(a_k) = \frac{k}{n(S)} = \frac{n(A)}{n(S)},$$

again as in Section 12.1.

Example 1. If two dice are rolled, what is the sample space? What are the probabilities of getting a total score of 2, 3, 4, 7?

Solution. In order to visualize the sample space containing all basic possible outcomes, consider the two dice to be distinguishable. Then denote each basic possible outcome as an ordered pair of numbers, where the first number represents die No. 1 and the second represents die No. 2. The sample space is

$$S = \{(1,1), (1,2), (1,3), \dots, (1,6), (2,1), (2,2), \dots, (6,6)\},$$

containing 36 equally likely basic possible outcomes.

(The ordered pair (h, t) has been abbreviated ht. But it would *not* be appropriate to abbreviate a pair of numbers such as (1, 2) by 12.)

Let A_k designate the event of "getting a total score of k." Then

$$A_2 = \{(1,1)\},$$

$$A_3 = \{(1,2), (2,1)\},$$

$$A_4 = \{(1,3), (2,2), (3,1)\},$$

and

$$A_7 = \{(1,6), (2,5), (3,4), (4,3), (5,2), (6,1)\}.$$

Thus

$$P(A_2) = \tfrac{1}{36}, \quad P(A_3) = \tfrac{2}{36} = \tfrac{1}{18}, \quad P(A_4) = \tfrac{3}{36} = \tfrac{1}{12}, \quad P(A_7) = \tfrac{6}{36} = \tfrac{1}{6}.$$

Note that among the possible totals (on the two dice) considered in Example 1, the most probable is 7. In Problem 2, you are asked to compute the probabilities of the other possible totals.

Rule (4) has the following important special case. Let A be any event in some sample space S, and let A^c be the complement of A. Thus $A \cap A^c = \emptyset$ and $A \cup A^c = S$. In other words, A and A^c are mutually exclusive events whose union is S. Then

$$P(A) + P(A^c) = P(S) = 1.$$

Example 2. If two dice are rolled, what is the probability that the total score is at least 4?

Solution. Consider this as a continuation of Example 1.

Let x be the total score on any roll of the two dice, and let A be the event $x \geq 4$. Then A^c is the event $x \leq 3$, i.e., $A^c = A_2 \cup A_3$. (Confirm this.)

Of course A_2 and A_3 are mutually exclusive events. So, using results from Example 1,

$$P(A^c) = P(A_2) + P(A_3) = 1/36 + 1/18 = 1/12.$$

Hence

$$P(x \geq 4) = P(A) = 1 - P(A^c) = 1 - 1/12 = 11/12.$$

The indirect approach of finding $P(A)$ by first computing $P(A^c)$ was used in Example 2 because $P(A^c)$ could be computed more easily. This approach will be useful in several future examples and problems.

Example 3. If two cards are drawn from a well-shuffled standard deck, what are the probabilities that (a) both are black? (b) both are red? (c) one is black and one is red?

Solution. The standard deck has 52 distinct cards. So the number of ways of drawing two cards in order is $52 \cdot 51$. (Look back to Section 11.2 if necessary.) Let S be the sample space of all possible ordered pairs. Then $n(S) = 52 \cdot 51$.

(a) Let B be the event that both cards are black. Then, since there are 26 black cards, $n(B) = 26 \cdot 25$. So

$$P(B) = \frac{n(B)}{n(S)} = \frac{26 \cdot 25}{52 \cdot 51} = \frac{25}{102}.$$

(b) Let R be the event that both cards are red. Then an analogous calculation gives $P(R) = 25/102$.

(c) Let A be the event that both cards are the same color. Thus $A = B \cup R$ and, since B and R are mutually exclusive events,

$$P(A) = P(B) + P(R) = 25/51.$$

Now it is easy to find the probability of the event that the two cards are different colors, for that is exactly the event A^c; and

$$P(A^c) = 1 - P(A) = 1 - 25/51 = 26/51.$$

The following valuable special case of Rule (4) generalizes the equation $P(A) + P(A^c) = 1$.

Corollary. *If A and B are any two events in a sample space S, then*

$$P(A \cap B) + P(A^c \cap B) = P(B). \tag{4'}$$

PROOF. Note that

$$(A \cap B) \cap (A^c \cap B) = \emptyset,$$

because no point can be in the left-hand side unless it belongs to both A and A^c. Also

$$(A \cap B) \cup (A^c \cap B) = B$$

because the points in the left-hand side consists of those points of B which belong to A together with those points of B which do not belong to A, i.e., all points of B.

Now (4′) follows at once from Rule (4). □

Example 4. Statistics indicate that approximately 1 out of 1500 adults in the United States will die of lung cancer within the next year and that 1 out of 2000 both smoke and will die of lung cancer within the next year. Find the probability that an adult in the United States does not smoke and will die of lung cancer within the next year.

Solution. For an adult in the United States selected at random, let A be the event "he or she smokes," and let B be the event "he or she will die of lung cancer within the next year." Then $P(B) = 1/1500 = 0.00067$ and $P(A \cap B) = 1/2000 = 0.0005$.

The objective is to find $P(A^c \cap B)$. Using (4′),

$$P(A^c \cap B) = P(B) - P(A \cap B) = 0.00067 - 0.0005 = 0.00017.$$

This example neither uses nor reveals the value of $P(A)$ itself. The statistics in Example 4 will be considered further in Section 12.6.

Problems

1. If a 25-year-old male has probability 0.0008 of dying in the next year, what is his probability of living the year?

2. If two dice are rolled, (a) what are the probabilities of getting a total score of 2, 3, 4, 5, 6, 7, 8, 9, 10, 11, and 12? (b) Do these probabilities add up to 1? (c) Which score(s) is (are) most probable? [This question is partly answered by Example 1.]

3. If two dice are rolled, what is the probability that the total score will be (a) 5 or less? (b) 6 or more?

4. If two cards are drawn from a (shuffled) standard deck, what is the probability that at least one is red?

5. If two dice are rolled, what are the probabilities that (a) both show 6? (b) neither shows 6? (c) at least one shows 6? (d) exactly one shows 6? [*Hint.* First decide which groups of these numbers, if any, ought to add up to 1.]

6. If a card is drawn from a standard deck, why is $P(\text{jack or spade}) \neq P(\text{jack}) + P(\text{spade})$?

7. If two cards are drawn from a standard deck, what is the probability that both are aces?

8. A die is weighted so that the probability of getting 6 is $\frac{1}{4}$ and the probabilities of the other five possible outcomes are equal. Find the probability of getting 1.

9. (a) How many of all the possible poker hands (five cards) will have no ace? (b) What is the probability of receiving a poker hand with no aces? (c) What is the probability of receiving a poker hand with at least one ace? [*Hint.* Refer to Example 5 of Section 11.2.]

10. What is the probability of receiving a poker hand containing all four aces?

11. If A and B are two events with $P(A) = \frac{1}{4}$ and $P(A \cap B) = \frac{1}{6}$, find $P(A \cap B^c)$.

12.3. The Basic Rules

The rules of probability encountered thus far are as follows.

For any event A in the sample space S of a certain experiment

$$0 \leq P(A) \leq 1. \tag{1}$$

$$P(S) = 1. \tag{2}$$

$$P(\varnothing) = 0. \tag{3}$$

For mutually exclusive events $A_1, A_2, \ldots, A_k$,

$$P(A_1 \cup A_2 \cup \cdots \cup A_k) = P(A_1) + P(A_2) + \cdots + P(A_k). \tag{4}$$

The next two examples introduce cases in which multiplication of probabilities is required.

Example 1. If a coin is tossed twice, what is the probability of getting two heads?

Solution. This is equivalent to an earlier example on the probability of two heads when *two* coins are tossed.

The sample space can be expressed as {hh, ht, th, tt}, where now the first letter in each ordered pair represents the first toss and the second letter represents the second toss. Since the four basic possible outcomes are equally likely, it follows at once that $P(\text{hh}) = \frac{1}{4}$.

But here is another point of view:

The coin, when tossed for the second time, has no memory of what it did before. So, for each toss, $P(\text{h}) = \frac{1}{2}$.

Now if the experiment (two tosses) is repeated many times, then on approximately half of those occasions the first toss yields heads. And of these occasions, approximately half yield heads on the second toss also. Thus

$$P(\text{hh}) = P(\text{h on 1st toss}) \cdot P(\text{h on 2nd toss}) = \tfrac{1}{2} \cdot \tfrac{1}{2} = \tfrac{1}{4}.$$

Example 2. If two cards are drawn from a well-shuffled standard deck, what is the probability of getting two aces?

Solution. This question is also familiar and it too can be answered by the methods of the previous sections. There are $52 \cdot 51$ ways of drawing an *ordered* pair. So the sample space S consists of $n(S) = 52 \cdot 51$ equally likely possible ordered pairs. Similarly, there are $4 \cdot 3$ ways of drawing two aces as an ordered pair. So in the sample space S, the event "two aces" contains $4 \cdot 3$ members. Hence, the probability of drawing two aces is

$$P(\text{two aces}) = \frac{n(\text{two aces})}{n(S)} = \frac{4 \cdot 3}{52 \cdot 51} = \frac{1}{221}.$$

Now consider the same question from another point of view.

The probability that the first card is an ace is $4/52 = 1/13$. So, if the experiment of drawing two cards is repeated many times, on approximately 1/13 of those occasions the first card should be an ace. And, on those occasions, the remainder of the deck will be 51 cards, among which are 3 aces. So the probability of drawing another ace is now $3/51 = 1/17$. That is, two aces will be drawn on approximately 1/17 of the occasions when the first card was an ace. Thus

$$P(\text{two aces}) = \frac{1}{13} \cdot \frac{1}{17} = \frac{1}{221}$$

$$= P(\text{1st card is ace}) \cdot P(\text{2nd card is ace if 1st is ace}).$$

For each of these examples, introduce two events.

In Example 1, let A be "h on first toss" and let B be "h on second toss."

In Example 2, let A be "first card is ace" and let B be "second card is ace."

In each example, the question was, what is the probability of the occurrence of both A and B, i.e., of the event $A \cap B$? And the result in each case can be expressed as

$$P(A \cap B) = P(A) \cdot P(B \text{ if } A \text{ occurs}).$$

But the two examples do differ.

In Example 1, the probability of heads on the second toss is $\frac{1}{2}$—*independent* of what happened on the first toss. Thus

$$P(B \text{ if } A \text{ has occurred}) \text{ is the same as } P(B).$$

While, in Example 2, the two drawings are *not independent*. That is,

$$P(\text{2nd card is ace if 1st is ace})$$

definitely depends on the fact that one ace has already been removed. In contrast to the coin which cannot remember its first toss, the deck of cards does "remember" what happened when the first card was removed.

Let us elevate the above to the status of another general rule. Mathematicians actually regard it as the *definition* of $P(B$ if A occurs), as will be emphasized in Section 12.6.

Rule. *The probability of events A and B both occurring is*

$$P(A \cap B) = P(A) \cdot P(B \text{ if } A \text{ occurs}). \tag{5}$$

Similarly, interchanging the roles of A and B,

$$P(A \cap B) = P(B) \cdot P(A \text{ if } B \text{ occurs}).$$

Note that in the special case when B is *independent* of A (5) simplifies to

$$P(A \cap B) = P(A) \cdot P(B). \tag{5'}$$

(The latter was illustrated by the coin-tossing example.)

The remaining examples and problems in this section are to be solved with the aid of Rules (1) through (5).

Remember that Rule (4)—the addition of probabilities—is used only for mutually exclusive events. On the other hand, Rule (5)—the multiplication of probabilities—is for cases when both events can occur simultaneously.

Example 3. What is the probability of getting two aces in two draws from a standard deck of cards if the first card drawn is replaced and the deck is reshuffled before the second drawing?

Solution. This is different from Example 2 inasmuch as the two drawings now *are independent.* (Replacing the first card and reshuffling the deck destroys the deck's memory of the first drawing.) Each drawn card now has the same probability of being an ace, namely, $4/52 = 1/13$. Thus by Rule (5), or (5′),

$$\begin{aligned} P(\text{two aces}) &= P(\text{1st card is ace}) \cdot P(\text{2nd card is ace}) \\ &= \frac{1}{13} \cdot \frac{1}{13} = \frac{1}{169}. \end{aligned}$$

Example 4. If two cards are drawn from a well-shuffled standard deck (without replacing the first card) what is the probability of getting the ace of spades and the ace of hearts?

Solution. There are two ways of producing this result. The first card might be the ace of spades and the second the ace of hearts *or* the first might be the ace of hearts and the second the ace of spades.

For the first case, note that

$$P(\text{1st card is ace of spades}) = \frac{1}{52}.$$

If the ace of spades has been drawn, the deck now contains 51 cards, exactly one of which is the ace of hearts. So now the probability of drawing the ace of hearts is $1/51$. It follows from Rule (5) that

$$\begin{aligned} &P(\text{1st card is ace of spades and 2nd is ace of hearts}) \\ &= P(\text{1st card is ace of spades}) \cdot P(\text{2nd is ace of hearts if 1st is ace of spades}) \\ &= \frac{1}{52} \cdot \frac{1}{51}. \end{aligned}$$

An analogous calculation yields

$$P(\text{1st card is ace of hearts and 2nd is ace of spades}) = \frac{1}{52} \cdot \frac{1}{51}.$$

The two cases are mutually exclusive, so Rule (4) gives

$$
\begin{aligned}
&P(\text{ace of spades and ace of hearts})\\
&= P(\text{1st card is ace of spades and 2nd is ace of hearts})\\
&\quad + P(\text{1st card is ace of hearts and 2nd is ace of spades})\\
&= \frac{1}{52}\cdot\frac{1}{51} + \frac{1}{52}\cdot\frac{1}{51} = \frac{1}{1326}.
\end{aligned}
$$

As one would expect, this is much less than the probability of getting *any* two aces as computed in Example 2.

Example 5. A packet of 10 outdated seeds contains 6 which will germinate and 4 which will not. But you cannot tell which are which. If you select two seeds from the package, what are the probabilities that (a) both seeds are good? (b) at least one is bad? (c) exactly one is bad?

Solution. (a) Note that

$$P(\text{1st seed selected is good}) = \frac{6}{10}.$$

Now if the first seed was good, there remain nine seeds among which five are good. Thus Rule (5) yields

$$
\begin{aligned}
P(\text{both seeds are good}) &= P(\text{1st is good}) \cdot P(\text{2nd is good if 1st is good})\\
&= \frac{6}{10}\cdot\frac{5}{9} = \frac{1}{3} = 0.33.
\end{aligned}
$$

(b) Let A be the event that both seeds are good. Then the event that at least one is bad is A^c. So, invoking Rule (4),

$$P(\text{at least one seed is bad}) = 1 - P(A) = 1 - 0.33 = 0.67.$$

(c) The event that one seed is good and the other is bad can occur in two different ways (compare with Example 4). Perhaps the first seed selected is good and the second is bad, or perhaps the first is bad and the second is good.

As in part (a), $P(\text{1st is good}) = 6/10$. Then, if a good seed has been removed, there remain nine seeds among which four are bad. So $P(\text{2nd is bad if 1st is good}) = 4/9$. Rule (5) then gives

$$
\begin{aligned}
P(\text{1st good and 2nd bad}) &= P(\text{1st good}) \cdot P(\text{2nd bad if 1st is good})\\
&= \frac{6}{10}\cdot\frac{4}{9} = \frac{4}{15}.
\end{aligned}
$$

Similarly

$$
\begin{aligned}
P(\text{1st bad and 2nd good}) &= P(\text{1st bad}) \cdot P(\text{2nd good if 1st bad})\\
&= \frac{4}{10}\cdot\frac{6}{9} = \frac{4}{15}.
\end{aligned}
$$

Then, since the two cases are mutually exclusive,

$$
\begin{aligned}
&P(\text{exactly one seed is bad}) \\
&\qquad = P(\text{1st good and 2nd bad}) + P(\text{1st bad and 2nd good}) \\
&\qquad = \frac{4}{15} + \frac{4}{15} = \frac{8}{15}.
\end{aligned}
$$

As a check on the above calculations, note that

$$
\begin{aligned}
P(\text{both seeds are bad}) &= P(\text{1st is bad}) \cdot P(\text{2nd is bad if 1st is bad}) \\
&= \frac{4}{10} \cdot \frac{3}{9} = \frac{2}{15}.
\end{aligned}
$$

Now the events "both seeds good," "exactly one bad," and "both seeds bad" are mutually exclusive and cover all possibilities. So their probabilities should add up to 1. See if they do.

Example 6. From the packet of six good seeds and four bad seeds, you now select three at random. Find the probabilities that (a) all three are good, (b) at least one is bad, (c) exactly one is good.

Solution. (a) As in Example 5,

$$
\begin{aligned}
P(\text{1st and 2nd seeds good}) &= P(\text{1st good}) \cdot P(\text{2nd good if 1st is good}) \\
&= \frac{6}{10} \cdot \frac{5}{9}.
\end{aligned}
$$

Now, if two good seeds have been selected, there remain eight seeds among which four are good. So

$$
\begin{aligned}
P(\text{all 3 good}) &= P(\text{1st and 2nd good}) \cdot P(\text{3rd good if 1st and 2nd good}) \\
&= \left(\frac{6}{10} \cdot \frac{5}{9}\right)\frac{4}{8} = \frac{1}{6} = 0.17.
\end{aligned}
$$

(b) Let A be the event that all three seeds are good. Then the event that at least one is bad is A^c. So, invoking Rule (4),

$$P(\text{at least one seed is bad}) = 1 - P(A) = 1 - 0.17 = 0.83.$$

(c) The event that exactly one seed is good is the union of three subcases. One possibility is that the first seed is good and the other two are bad. The probability of this event is found as follows.

$$P(\text{1st seed is good}) = \frac{6}{10}.$$

After a good seed has been removed, there remain nine seeds among which four are bad. So

$$
\begin{aligned}
&P(\text{1st seed good and 2nd bad})\\
&\quad = P(\text{1st seed good}) \cdot P(\text{2nd bad if 1st is good})\\
&\quad = \frac{6}{10} \cdot \frac{4}{9} = \frac{4}{15}.
\end{aligned}
$$

After one good seed and one bad are removed the package contains eight seeds among which three are bad. So

$$
\begin{aligned}
&P(\text{1st seed is good and 2nd and 3rd are bad})\\
&\quad = P(\text{1st good and 2nd bad}) \cdot P(\text{3rd bad if 1st good and 2nd bad})\\
&\quad = \left(\frac{6}{10} \cdot \frac{4}{9}\right)\frac{3}{8}.
\end{aligned}
$$

In a similar manner you should show that

$$P(\text{2nd seed is good while 1st and 3rd are bad}) = \left(\frac{4}{10} \cdot \frac{6}{9}\right)\frac{3}{8},$$

and

$$P(\text{3rd seed is good while 1st and 2nd are bad}) = \left(\frac{4}{10} \cdot \frac{3}{9}\right)\frac{6}{8}.$$

The three cases considered above are mutually exclusive and cover all possible ways to select one good and two bad seeds. Thus

$$P(\text{one good seed and two bad}) = \frac{6}{10} \cdot \frac{4}{9} \cdot \frac{3}{8} + \frac{4}{10} \cdot \frac{6}{9} \cdot \frac{3}{8} + \frac{4}{10} \cdot \frac{3}{9} \cdot \frac{6}{8} = \frac{3}{10} = 0.3.$$

Further examples similar to this will arise in considerations of "quality control" in Section 12.4.

Problems

1. The chains produced by a certain factory are rated at 5000 pounds breaking strength. But a chain is "no stronger than its weakest link," and 1 link in 100 fails to satisfy the advertised strength rating. Find the probabilities that a length of this chain having 40 links (a) will satisfy the 5000-pound rating, (b) will actually break below the 5000-pound point. [*Hint*. When this problem is properly interpreted you will need to compute $(0.99)^{40}$. This can be done quite efficiently with even a simple pocket calculator. Note that $r^{40} = (r^8)^4 r^8$. Most calculators can easily compute squares. So find r^8 by three squarings and store the result in the calculator's "memory." Then compute $(r^8)^4$ by two more squarings and multiply by r^8 from the memory. This idea appeared earlier in Section 6.3.]

2. Explain why each of the following statements is *false*.
   (a) If a *fair* die has failed to produce a 5 even once in 100 successive rolls, it is long overdue for a 5 (in order that approximately one-sixth of the rolls yield 5). So

the probability of 5 on the 101st roll will be greater than $\frac{1}{6}$. This is an example of the "Monte Carlo fallacy"—the downfall of many gamblers.

(b) If a single card is drawn from a standard deck, the probability that it is a jack is 1/13, and the probability that it is a diamond is $\frac{1}{4}$. So, by Rule (4), the probability that it is either a jack or a diamond is $1/13 + 1/4$.

(c) When two coins are tossed $P(\text{hh}) = \frac{1}{4}$ and $P(\text{tt}) = \frac{1}{4}$. So, by Rule (5), P(two heads and two tails) $= P(\text{hh}) \cdot P(\text{tt}) = 1/16$.

3. What is the probability that at least one child in a family is a girl if there are (a) three children in the family?, (b) four children in the family?

4. What is the probability of having three girls and a boy (in any order) in a family of four children?

5. Seeds in a certain batch have an 80% germination rate, i.e., each seed has probability 0.8 of germinating. So you plant two seeds in the same pot. (a) What is the probability that at least one will germinate. (b) What is the probability that (exactly) one will germinate?

6. Seeds in a certain batch have only a 50% germination rate. If you plant three seeds in one pot, what is the probability that at least one will germinate?

7. In four rolls of a die, what is the probability of getting at least one 6? (This and Problem 8 are two of the questions which gambler Chevalier de Mere posed to Blaise Pascal in 1654, and which led to a systematic study of probability theory by Pascal and Pierre de Fermat.)

8. In 24 rolls of a pair of dice, what is the probability of getting a score of 12 at least once?

9. Four steel bolts are selected blindly from a box containing 19 good ones and 5 with defective threads. Find the probabilities that (a) at least one of the four is defective, (b) exactly one of the four is defective, (c) at most one of the four is defective.

10. From informal observations of traffic on the Newport Bridge in Rhode Island, one might conclude that the probability of breakdown for an "average, well-prepared" automobile traveling $\frac{1}{4}$ mile under good conditions is 0.00001. (a) Show that the probability of 4000 such vehicles completing a $\frac{1}{4}$-mile trip under good conditions without any breakdowns is 0.9608. [*Hint.* You will need to compute $(0.99999)^{4000}$. To find r^{4000} fairly efficiently use the fact that

$$r^{4000} = (r^{32})^{125} = (r^{32})^{128}/(r^{32})^3.$$

Now compute r^{32}, via repeated squarings. Then store this in the "memory" for use in the rest of the computation.]

(b) What is the probability of at least one breakdown if 4000 such automobiles set forth on a trip of $\frac{1}{4}$ mile under good conditions?

11. In case of an anticipated nuclear "exchange," the Civil Defense "Crisis Relocation Plans" for Rhode Island would have about 200,000 cars traveling north on U.S. 95 and 495 to "low-risk areas" in New Hampshire. (a) What is the probability of at least one breakdown in any given quarter mile? [*Hint.* $r^{200,000} = (r^{64})^{3125}$. Now set $s = r^{64}$, and note that $s^{3125} = (s^{1024}s^{16})^3 s^5$.] (b) What is the probability of at least one breakdown in any given mile? (c) What effect would such breakdowns

have on the rest of the traffic? (Consider, in particular, a breakdown—or accident—at an "on ramp.")

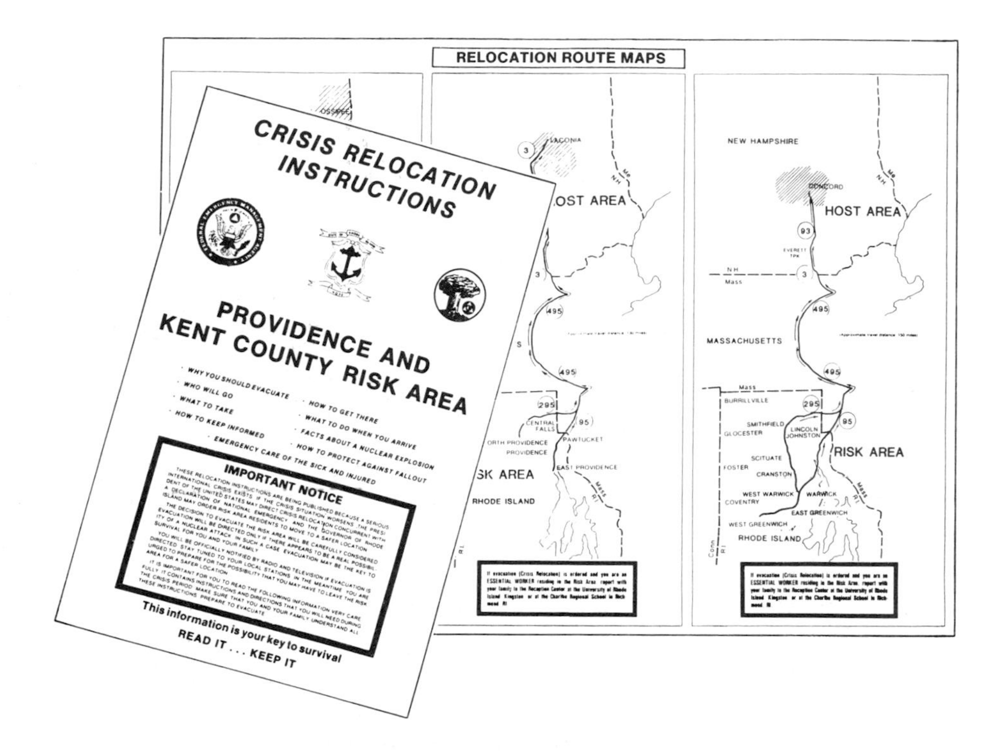

12. If three cards are drawn (without replacement) from a standard well-shuffled deck (a) what is the probability of getting three aces? (b) What is the probability of getting the ace of spades, the ace of hearts, and the ace of diamonds? [*Hint*. For (b) try to follow the method of Examples 4 and 6(c), noting that there are 3! different orders in which one might draw the three required aces. Try to check your answer another way.]

12.4. Quality Control (Optional)

One of the problems in the production of goods (in a factory or on a farm) is "quality control." To be certain that the output from a production process meets specified quality standards, one would have to inspect or test each individual item. Indeed this is done for high-priced durable goods such as automobiles and major appliances.

But in many cases it is impractical to check each item. So instead, one inspects or tests only a small sample of the output, and then assumes that the quality of the sampled items represents the quality of the entire batch from which the sample was selected.

For example, if you were mass producing a low-cost item by machine, the cost of individually testing each item might exceed the cost of production—making such complete testing unrealistic.

And in some cases testing actually destroys or damages the item being tested: To determine the tensile strength of a sample of string one gradually applies more and more tension until the string breaks. To test a razor blade for sharpness, one cuts something with it—after which it can no longer be sold as a new blade. To determine the point at which an electrical fuse will "blow," one gradually increases the electrical current through the fuse until it does blow. And to determine whether an orange crop is ready for harvest, the grower has to eat some oranges.

The more items tested and found satisfactory from a batch, the more likely that the batch is all right. But also the more tested, the higher the cost of testing.

The examples below show how to calculate the probability of finding defective products by testing a sample. Then, in real life, someone also has to decide what risk of defective products escaping detection is an acceptable risk.

Example 1. A mass production process is turning out semiconductor "chips" for use in computers. A batch of 40 of these chips contains 5 which are defective. If an inspector tests just two chips taken blindly from this batch, what is the probability that at least one will be defective?

Solution. First compute the probability that both chips tested are OK. If A is the event "both chips are OK," then the event "at least one is defective" is A^c, and $P(A^c) = 1 - P(A)$.

For the first chip selected

$$P(\text{1st chip is OK}) = \frac{35}{40}.$$

Now if the first chip was good, there remain 39 chips among which 34 are good. Thus

$$P(\text{2nd chip is OK if 1st is OK}) = \frac{34}{39}.$$

So, by Rule (5),

$$\begin{aligned} P(A) &= P(\text{both chips are OK}) \\ &= P(\text{1st chip is Ok}) \cdot P(\text{2nd chip OK if 1st is OK}) = \frac{35}{40} \cdot \frac{34}{39} = 0.76. \end{aligned}$$

From this

$$P(\text{at least one of the two chips is defective})$$
$$= P(A^c) = 1 - P(A) = 1 - 0.76 = 0.24.$$

Thus, the inspector is not likely to find even 1 of the 5 defective chips if he tests only 2 out of a batch of 40.

Example 2. A batch of 400 semiconductor chips contains 50 which are defective. If an inspector tests just 20 chips selected at random from this batch, what is the probability that at least 1 will be defective? [The percentage of defective items in the batch and the percentage of the batch tested are the same as in Example 1. But the result will be very different.]

Solution. Again it is easier to approach the question indirectly, first computing $P(\text{all 20 chips tested are OK})$. This is found as in Example 1, but now with *repeated* use of Rule (5):

$$P(\text{1st chip tested is OK}) = \frac{350}{400}.$$

Assuming the first chip was OK, there remain 399 chips of which 349 are OK. Thus,

$$P(\text{1st and 2nd are OK}) = P(\text{1st is OK}).\ P(\text{2nd is OK if 1st is OK})$$
$$= \frac{350}{400} \cdot \frac{349}{399}.$$

Similarly,

$$P(\text{1st, 2nd, and 3rd are OK})$$
$$= P(\text{1st and 2nd OK}) \cdot P(\text{3rd OK if 1st and 2nd are OK})$$
$$= \left(\frac{350}{400} \cdot \frac{349}{399}\right) \frac{348}{398} = \frac{350}{400} \cdot \frac{349}{399} \cdot \frac{348}{398}.$$

Perhaps you can now see how the pattern continues. The result is

$$P(\text{all 20 are OK}) = \frac{350}{400} \cdot \frac{349}{399} \cdot \frac{348}{398} \cdots \frac{331}{381}.$$

The multiplication of these 20 fractions is straightforward but *tedious*—even with the aid of a calculator. However there is an easy way of getting a good approximation to the answer. Note that among the 20 fractions to be multiplied together, the largest is 350/400 and the smallest is 331/381. If you replaced each fraction by the smallest, the result would be a lower estimate for the true product. Similarly, if you replaced each fraction by the largest, the result would be an upper estimate for the true product. Thus

$$\left(\frac{331}{381}\right)^{20} < P(\text{all 20 are OK}) < \left(\frac{350}{400}\right)^{20}.$$

A method for computing powers such as r^{20} with just a simple calculator was indicated earlier. [Note that $r^{20} = (r^4)^4 r^4$.] You should find

$$0.06 < P(\text{all 20 are OK}) < 0.07.$$

To be conservative, just say $P(\text{all 20 are OK}) < 0.07$.

It then follows that

$$\begin{aligned} P(\text{at least one chip tested is defective}) &= 1 - P(\text{all 20 are OK}) \\ &> 1 - 0.07 = 0.93. \end{aligned}$$

Thus, assuming the same rate of defectives and the same percentage sampling as in Example 1, the use of much larger batches gives a high probability of finding at least one of the defective chips.

Example 3. If 20 chips are selected at random from the batch of 400 in Example 2, find the probabilities that (a) all 20 of these will be defective, (b) that (exactly) 1 of these will be defective.

Solution. (a) Analogously to the solution of Example 2,

$$P(\text{1st chip tested is defective}) = \frac{50}{400}.$$

Then,

$$\begin{aligned} &P(\text{1st and 2nd are defective}) \\ &= P(\text{1st is defective}) \cdot P(\text{2nd is defective if 1st is defective}) \\ &= \frac{50}{400} \cdot \frac{49}{399}. \end{aligned}$$

Continuing as in Example 2, you eventually find

$$P(\text{all 20 are defective}) = \frac{50}{400} \cdot \frac{49}{399} \cdot \frac{48}{398} \cdots \frac{31}{381} < \left(\frac{50}{400}\right)^{20} = 9 \times 10^{-19}.$$

Not surprisingly, the event that all 20 sampled items are defective is highly improbable.

(b) The probability of encountering exactly 1 defective chip among the 20 tested requires a different type of computation. You can encounter exactly one defective item in 20 different ways. You might find that the first is defective and the others are OK, or the second is defective and the others are OK, or the third is defective and the others are OK, etc. The probabilities for these 20 mutually exclusive subcases must be added together to find the probability that exactly 1 of the 20 chips is defective (regardless of which one it is).

To compute $P(\text{1st is defective, others OK})$, begin by noting that

$$P'(\text{1st is defective}) = \frac{50}{400}.$$

Assuming that the first was defective, there will remain in the batch 399 chips among which 350 are OK. So

$$\begin{aligned} &P(\text{1st is defective and 2nd is OK}) \\ &= P(\text{1st is defective}) \cdot P(\text{2nd is OK if 1st is defective}) \\ &= \frac{50}{400} \cdot \frac{350}{399}. \end{aligned}$$

Continuing one finds

$$P(\text{1st is defective and other 19 are OK}) = \frac{50}{400} \cdot \frac{350}{399} \cdot \frac{349}{398} \cdots \frac{332}{381}.$$

The value of this product lies between

$$\frac{50}{400}\left(\frac{332}{381}\right)^{19} = 0.009 \quad \text{and} \quad \frac{50}{400}\left(\frac{350}{399}\right)^{19} = 0.01.$$

Call it 0.01.

Now compute

$$\begin{aligned} &P(\text{1st is OK and 2nd is defective}) \\ &= P(\text{1st is OK}) \cdot P(\text{2nd is defective if 1st is OK}) \\ &= \frac{350}{400} \cdot \frac{50}{399}. \end{aligned}$$

Continuing as above,

$$P(\text{2nd is defective and all others are OK}) = \frac{350}{400} \cdot \frac{50}{399} \cdot \frac{349}{398} \cdots \frac{332}{351},$$

equivalent to the previous case. So again the value, rounded off, is 0.01.

In fact, all 20 cases have this same probability. Thus

$$\begin{aligned} P(\text{exactly one is defective}) &= P(\text{1st defective, others OK}) \\ &\quad + P(\text{2nd defective, others OK}) + \cdots \\ &\quad + P(\text{20th defective, others OK}) \\ &= 0.01 \times 20 = 0.2. \end{aligned}$$

In practice, the quality-control manager may decide to accept a batch of chips if the sample of 20 shows no more than 1 defective, and reject the batch if the sample yields 2 or more defectives.

Example 4. Under the same assumptions as in Example 2—a batch of 400 containing 12.5% defectives—what is the probability of having *at least 2* defective chips in the sample of 20 tested?

Solution. This is easy on the basis of the results already obtained. Rule (4) gives

$$P(\text{all OK}) + P(\text{exactly one defective}) + P(\text{at least two defective}) = 1,$$

since these are the probabilities of three mutually exclusive events covering all possible cases. Thus

$$\begin{aligned} P(\text{at least two defective}) &= 1 - P(\text{all OK}) - P(\text{exactly one defective}) \\ &\cong 1 - 0.07 - 0.2 = 0.73. \end{aligned}$$

Books devoted to probability theory provide systematic methods for also computing the probability of finding "exactly two" defectives, or "exactly three," etc., and hence the probability of "at least three," or "at least four," etc. Such calculations are beyond the goals of this brief introduction.

Some ideas for this section came from N. L. Enrick, *Quality Control and Reliability* (Industrial Press Inc., New York, 1972).

Problems

1. A quality-control inspector selects 5 out of each batch of 50 garden rakes for inspection. (a) If there are 10 rakes with defective welds in a batch of 50, what is the probability that the inspector will catch at least 1 of them? (b) The production manager is dissatisfied. When workmanship deteriorates to the extent that 10 defectives occur in a batch of 50, she wants to know with probability at least 85% (0.85), so that the problem can be corrected. How many items must the inspector examine from each batch of 50 to achieve this goal?

2. Ten bolts are selected at random for inspection out of a batch of 100 from a machine producing the bolts. If the batch contains 5% with defects (in the threads), what is the probability that (a) all those tested are good?, (b) at least one is defective?

3. A batch of 200 bolts with 5% defective is judged on the basis of a 10% sample (20 bolts). What is the probability that (a) all those tested are good?, (b) at least one is defective?

4. Now a batch of 500 bolts with 5% defective is judged on the basis of a 10% sample. What is the probability that (a) all those tested are good?, (b) at least one is defective?

5. The buttons produced by a certain button-making machine are collected in batches of 1000, and the quality-control inspector checks 50 buttons out of each batch. If a batch contains 4% defective buttons, find (a) P(all 50 are OK), (b) P(exactly one of those tested is defective). [*Hint.* $r^{50} = (r^{16})^2 r^{16} r^2$.]

6. The inspector in Problem 5 accepts a batch if the sample of 50 buttons examined contains no more than 1 defective. What is the probability that the inspector will accept a batch which contains 4% defective buttons?

7. What is the probability that the inspector in Problems 5 and 6 will accept a batch of buttons containing only 1% defectives?

8. What is the probability that the inspector in Problems 5 and 6 will accept a batch of buttons containing 3% defectives?

12.5. Expectation

The "expectation" of a quantity x will be defined below as the average value which would be obtained over repetitions of an experiment if the various possible values of x occurred exactly as often as the probabilities suggest.

Example 1. Let x be the number which comes up when a fair die is rolled. If the die is rolled m times, where m is a large number, one expects to get each of the six possible outcomes about $m/6$ times. Thus, the average score for the m rolls is expected to be about

$$\frac{1(m/6) + 2(m/6) + 3(m/6) + 4(m/6) + 5(m/6) + 6(m/6)}{m}$$

$$= \frac{1 + 2 + 3 + 4 + 5 + 6}{6} = 3.5.$$

Note that, since $P(x=k) = \frac{1}{6}$ for each outcome $k = 1, 2, 3, 4, 5$, and 6, the above can be written as

$$1P(x=1) + 2P(x=2) + 3P(x=3) + 4P(x=4) + 5P(x=5) + 6P(x=6).$$

This simple example motivates the following.

Definition. Let $a_1, a_2, \ldots$, and a_k be all the possible values of the result x from some experiment. Then the **expectation** or **expected value** of x is

$$E(x) = a_1 P(x=a_1) + a_2 P(x=a_2) + \cdots + a_k P(x=a_k).$$

This definition represents a "weighted average" of the possible values of x, with the probabilities providing the "weights."

Along with the definition of expectation, you will need to remember that

$$P(x=a_1) + P(x=a_2) + \cdots + P(x=a_k) = 1.$$

Example 2. Compute the expectation of the total score when two dice are rolled.

Solution. The possible values of x, the total score, are 2, 3, 4, 5, 6, 7, 8, 9, 10, 11, and 12, and their probabilities were to be computed in a problem in Section 12.2. Using these, one finds

$$\begin{aligned}E(x) &= 2P(x=2) + 3P(x=3) + \cdots + 12P(x=12)\\&= 2(1/36) + 3(2/36) + 4(3/36) + 5(4/36) + 6(5/36) + 7(6/36)\\&\quad + 8(5/36) + 9(4/36) + 10(3/36) + 11(2/36) + 12(1/36)\\&= 252/36 = 7.\end{aligned}$$

In Section 12.2, you found that 7 was the most probable score from a roll of two fair dice. It just so happens, for this particular example, that the same value is the expected average score in multiple repetitions of the experiment. This will not necessarily be the case for other examples.

Example 3. An outdated packet of seeds contains six which are still good and four which are not. If three seeds are selected at random from this package, what is the expectation for the number of good seeds?

Solution. You might guess that the answer is $6 \times 3/10 = 1.8$, and this is correct. But, for illustration, here is the detailed argument.

Let x be the number of good seeds among those chosen. Then x must be 0, 1, 2, or 3. Now, from Example 6 of Section 12.3, $P(x=3) = 1/6$ and $P(x=1) = 3/10$. Also it is easy to compute

$$P(x=0) = P(\text{all three are bad}) = \frac{4}{10}\cdot\frac{3}{9}\cdot\frac{2}{8} = \frac{1}{30}.$$

Now use the fact that $P(x=0) + P(x=1) + P(x=2) + P(x=3) = 1$ to find

$$P(x=2) = 1 - P(x=0) - P(x=1) - P(x=3) = 1 - \frac{1}{30} - \frac{3}{10} - \frac{1}{6} = \frac{1}{2}.$$

[For practice, confirm this result by computing $P(x=2)$ directly.]

Putting it all together, one gets

$$\begin{aligned}E(x) &= 0P(x=0) + 1P(x=1) + 2P(x=2) + 3P(x=3)\\&= 0\left(\frac{1}{30}\right) + 1\left(\frac{3}{10}\right) + 2\left(\frac{1}{2}\right) + 3\left(\frac{1}{6}\right) = \frac{18}{10} = 1.8.\end{aligned}$$

Example 4. Statistics indicate that the probability of death in the next year for the "average" 20-year-old male is 0.0012, and for the "average" 20-year-old female is 0.0006. How much should a life insurance company charge a 20 year old for a 1-year \$1000 death-benefit policy so that the company will have an expected gross profit of \$10 (a) if the subject is male, (b) if the subject is female, and (c) if the company is not allowed to distinguish males from females?

Solution. Let c be the charge for the policy in dollars, and let x be the gross profit (in dollars) for the company from such a policy. If the subject lives the entire year, then $x = c$; if the subject dies before the end of the year, then $x = c - 1000$ (which will be negative since certainly $c < 1000$). The company wants $E(x) = 10$.

So set

$$E(x) = cP(\text{subject lives}) + (c - 1000)\,P(\text{subject dies}) = 10.$$

(a) If the subject is male, this becomes

$$c(0.9988) + (c - 1000)(0.0012) = 10$$

or $c - 1.2 = 10$. So $c = 11.20$.

(b) If the subject is female, an analogous calculation gives $c = 10.60$. Check it.

(c) If the company must charge the same premium for both sexes, and *if* it expects to sell as many policies to females as to males, then it can assume the probability for death in the next year is the average of 0.0012 and 0.0006, namely, 0.0009. Then $c = 10.90$.

Example 5. Start with \$100 as your bankroll for betting on tosses of a fair coin. For each toss, you bet half the money you have at that time on "heads", and you win or lose the amount of your bet. What is the expected value of your net gain (or loss) after the coin has been tossed twice?

Solution. Note that when you win your bankroll is multiplied by $\frac{3}{2}$, and when you lose it is multiplied by $\frac{1}{2}$. (Confirm this.) So, for example, if you win once and lose once you will end up with

$$\$100(\tfrac{3}{2})(\tfrac{1}{2}) = \$100(\tfrac{3}{4}) = \$75.$$

(If you have studied Section 6.1, recall Problem 7 of that section.)

The sample space for two tosses of the coin is

$$S = \{\text{hh, ht, th, tt}\},$$

and each basic possible outcome has probability $\frac{1}{4}$. Note that in three of the four possible cases, you lose. The only time you come out ahead is in the event of two heads. But it does not necessarily follow that you must expect to lose money "on the average."

Your bankroll after two tosses has three possible values, namely,

$$\$100(\tfrac{3}{2})^2 = \$225, \quad \$100(\tfrac{3}{2})(\tfrac{1}{2}) = \$75, \quad \text{and} \quad \$100(\tfrac{1}{2})^2 = \$25.$$

Let x be your net gain in dollars in two tosses. (In case you lose, x will be a negative number.) You must have one of the following:

$$x = 225 - 100 = 125, \quad x = 75 - 100 = -25, \quad \text{or} \quad x = 25 - 100 = -75.$$

The probabilities of these three events are

$$P(x = 125) \;= P(\text{hh}) = \tfrac{1}{4},$$

$$P(x = -25) = P(\text{ht or th}) = \tfrac{1}{2},$$

and

$$P(x=-75) = P(\text{tt}) = \tfrac{1}{4}.$$

So the expectation for your net gain is

$$\begin{aligned} E(x) &= (125)P(x=125) + (-25)P(x=-25) + (-75)P(x=-75) \\ &= 125(\tfrac{1}{4}) - 25(\tfrac{1}{2}) - 75(\tfrac{1}{4}) = 0. \end{aligned}$$

Thus, while you are most likely to lose in any single two-toss game, you can expect that in many repetitions of this gambling experiment your average result would be to break even.

Definition. A game is **fair** if the expected value of the gain for each player is zero. The bets in a game or experiment are **fair** if they result in a fair game.

Example 6. If you bet 1 dollar that the roll of a fair die will produce a 4, what should the payoff be to make this a fair bet?

Solution. Let p be the payoff (in dollars) if you win, and let x be your gain on any particular roll of the die. Then either $x = p - 1$ or $x = -1$. The probabilities of these two events are

$$P(x=p-1) = \tfrac{1}{6} \qquad \text{and} \qquad P(x=-1) = \tfrac{5}{6}.$$

So

$$E(x) = (p-1)P(x=p-1) + (-1)P(x=-1) = (p-1)/6 - 5/6.$$

In order to make $E(x) = 0$—for a fair bet—one requires

$$p = 6.$$

One often talks of the odds in favor or the odds against some event.

Definition. The **odds in favor** of an event A are "r to s" where r and s are any two numbers such that

$$\frac{r}{s} = \frac{P(A)}{P(A^c)} = \frac{P(A)}{1 - P(A)},$$

and the **odds against** event A are "s to r."

Example 7. What are the odds against getting a 4 on the roll of a fair die?

Solution. Let A be the event of getting a 4. Then $P(A) = \tfrac{1}{6}$. So

$$\frac{P(A)}{P(A^c)} = \frac{1/6}{5/6} = \frac{1}{5},$$

and one says the odds are "five to one" against getting a 4. (It would also be correct to say the odds are "ten to two" or "thirty-five to seven" against getting a 4. But one usually prefers the simplest numbers.)

Even though a game may be fair, or almost fair, there is another problem facing anyone who intends to gamble against a much wealthier opponent, e.g., a gambling casino. If you run out of money to bet, you are finished. This is referred to as "gambler's ruin."

Example 8 (Gambler's Ruin). You have a bankroll of \$100 to gamble and your opponent has \$700. You repeatedly make 1 dollar, even-money bets on the outcome of tossing a fair coin. So for each toss you either win a dollar or lose a dollar in this fair game. The game continues until either you or your opponent is "ruined." What is the probability that you will win your opponent's \$700 before he or she wins your \$100?

Solution. There is no way of predicting how long the play will continue, but assume that it does not continue forever. In other words, someone is ruined after a finite number of plays.

Let x be your net gain (in dollars) when the game ends. Thus, either $x = -100$ or $x = 700$. Since the coin is fair, $E(x)$ must be zero. That is, you should break even "on the average" if the experiment is repeated many times.

Since the event "$x = -100$" is the complement of the event "$x = 700$," $P(x=-100) = 1 - P(x=700)$. Substitute this into the requirement

$$E(x) = (-100)P(x=-100) + (700)P(x=700) = 0$$

to get

$$-100[1 - P(x=700)] + (700)P(x=700) = 0.$$

Solving this, one finds

$$P(x=700) = \tfrac{1}{8}.$$

So, the player with the smaller bankroll is likely to lose it. (The odds in this case are 7 to 1 against you. Check that statement.)

In the previous examples of this section, probabilities were used to compute expectations. But in Example 8, the known expectation was used to find a probability.

Problems

1. A fair coin is tossed once. Let $x = 1$ if it comes up heads, and let $x = 0$ if it comes up tails. Find $E(x)$.
2. A fair die is rolled and the score x is taken to be the square of the number which comes up. Find $E(x)$.
3. A die is weighted so that the probability of getting 6 is $\frac{1}{4}$, and the probabilities of the other five possible outcomes are equal. If x is the number which comes up on any roll of the die, find (a) $P(x=1)$, (b) $E(x)$.

4. If you bet 1 dollar that the card you draw from a well-shuffled standard deck will be an ace, (a) what should the payoff be in order that this be a fair bet? (b) What are the odds against your success?

5. Seeds in a certain batch have an 80% germination rate. If you plant two in a pot and let x be the number which germinate, what are (a) $P(x=0)$, (b) $P(x=1)$, (c) $P(x=2)$, (d) $E(x)$?

6. A roulette wheel has 18 red slots, 18 black slots, and 2 green slots. If you bet 1 dollar on red and the ball stops in a red slot, you win 1 dollar; if it stops elsewhere you lose. What is your expected gain in each play?

7. If a life-insurance company charges $100 for a 1-year death-benefit policy of $10,000 for a 25 year old, what is the expected gross revenue to the company from that policy? Assume that the subject has probability 0.0008 of dying during the year.

8. You have $25 to gamble with and your opponent has $50. You make repeated even-money fair bets on the toss of a coin until one of you is ruined. What is your probability of getting your opponent's $50 before he or she gets your $25?

9. A one-dollar lottery ticket offers a chance at a grand prize of $100,000 and a second prize of $10,000. If you bought one of the 150,000 tickets sold, (a) what is the expectation of your gain? (b) What are the odds against your winning anything?

10. If the odds are 17 to one against event A, what is $P(A)$?

11. Rework Example 5 for three tosses of a fair coin.

*12. Rework Example 5 for four tosses of a fair coin, and show that $E(x) = 0$ once again. This suggests that there really is no paradox in Problem 7 of Section 6.1. However, it would be difficult at this stage to compute $E(x)$ for the case of 10 even-money fair bets.

12.6. Conditional Probability

Partial knowledge about an "experiment" can affect the probabilities of the events of interest.

For example, suppose the experiment under consideration consists of two tosses of a fair coin, and the event of interest is "two heads." If that is *it*, then the probability of the event is $\frac{1}{4}$. However, if we have information that the first toss of the coin has already been made and it came up tails, then the probability of two heads drops to zero.

The concept of the probability of a certain event under the condition that some other event occurs (or has occurred) was encountered, although not named, in Rule (5),

$$P(A \text{ and } B) = P(A \cap B) = P(A) \cdot P(B \text{ if } A \text{ occurs}). \tag{5}$$

The expression $P(B \text{ if } A \text{ occurs})$ is called a "conditional probability" and represents the probability of event B under the condition that event A occurs. Interchanging the roles of A and B yields the companion equation

$$P(A \cap B) = P(B) \cdot P(A \text{ if } B \text{ occurs}).$$

Example 1. What is the probability of getting 13 heads in 13 tosses of a fair coin (a) if no other information is available? (b) if the first 12 tosses have already been completed and have produced heads each time?

Solution. (a) Since the 13 tosses are independent, Rule (5′) applied repeatedly gives

$$P(13 \text{ heads}) = \left(\frac{1}{2}\right)^{13} = \frac{1}{8192}.$$

(b) After witnessing 12 heads in succession, many gamblers would want to "bet on tails" next because they know that the probability of 13 heads in a row is very small. This is an example of the "Monte Carlo fallacy" again. The reasoning is fallacious because each toss of the coin is independent. (The coin has no memory.) Thus the probability of heads on the thirteenth toss is still $\frac{1}{2}$, even after a run of 12 heads. So

$$P(13 \text{ heads if first 12 are heads}) = \tfrac{1}{2}.$$

Example 2. Rework Example 1(b) making use of Rule (5): $P(13 \text{ heads}) = P(\text{first 12 are heads}) \cdot P(13 \text{ heads if first 12 are heads})$.

Solution. From above, $P(13 \text{ heads}) = (\frac{1}{2})^{13}$. And, similarly, $P(\text{first 12 are heads}) = (\frac{1}{2})^{12}$. So Rule (5) gives

$$(\tfrac{1}{2})^{13} = (\tfrac{1}{2})^{12}\, P(13 \text{ heads if first 12 are heads}).$$

Solving this for the conditional probability yields (once again)

$$P(13 \text{ heads if first 12 are heads}) = \tfrac{1}{2}.$$

The Monte Carlo fallacy sometimes arises in speculation about the sex of a forthcoming baby. If the family already has three boys, one is tempted to think that the forthcoming baby is more likely to be a girl since the probability of *four* boys is only $\frac{1}{16}$ (assuming the two sexes are equally likely at birth). But (as in Example 1) the probability of four boys *if the first three children born are boys* is $\frac{1}{2}$—not $\frac{1}{16}$. In other words, the sex of a forthcoming baby is not influenced by the sexes of its siblings.

In Example 1, it was taken on faith that a coin has no memory, so that each toss is independent of the ones which preceded it. And in the comments above it was asserted that the sex of a baby is independent of the sexes of its siblings.

In less apparent situations, a mathematician would say that one *determines* whether two events, A and B, are "independent" by examining the probabilities $P(A)$, $P(B)$, and $P(A \cap B)$. The mathematician actually *defines* independence as follows.

Definition. Events A and B are **independent** if and only if

$$P(A \cap B) = P(A) \cdot P(B). \tag{5'}$$

Definition. Whether or not A and B are independent, the **conditional probabilities** are defined, using Rule (5), by

$$P(A \text{ if } B \text{ occurs}) = \frac{P(A \cap B)}{P(B)} \qquad \text{assuming } P(B) \neq 0$$

and

$$P(B \text{ if } A \text{ occurs}) = \frac{P(A \cap B)}{P(A)} \qquad \text{assuming } P(A) \neq 0.$$

Here is a problem which can be resolved either with or without the definition of conditional probability. Considering it both ways should increase one's confidence in the definition. Not until Example 5 will you encounter a problem that *requires* the use of the definition.

Example 3. A woman has two cats, one grey and the other black. (a) What is the probability that both are males? (b) Suppose a visitor asks if one is male, and the owner says yes. Then what is the probability that both are males? (c) Suppose the visitor asks if the grey cat is male, and the owner says yes. Now what is the probability that both are males?

Solution. Here, no use will be made of the definition of conditional probability.

(a) The sample space of basic possible outcomes can be represented as $S = \{\text{mm, mf, fm, ff}\}$, where the sex of the grey cat is listed first. With no auxiliary information, the four members of the sample space are equally likely. So

$$P(\text{mm}) = \tfrac{1}{4}.$$

(b) Now, armed with the knowledge that at least one of the cats is male, the visitor has a new sample space $\{\text{mm, mf, fm}\}$. And, as far as the visitor is concerned, each member is equally likely. So

$$P(\text{mm if one cat is m}) = \tfrac{1}{3}.$$

(c) Finally, if the visitor knows that the grey cat is male, the sample space becomes $\{\text{mm, mf}\}$, and, since these are equally likely (from the visitor's viewpoint),

$$P(\text{mm if grey cat is m}) = \tfrac{1}{2}.$$

Example 4. Rework parts (b) and (c) of Example 3 using the definition of conditional probability.

Solution. (b) Note that $P(\text{mm}) = P(\text{mm} \cap \text{“one cat is m”})$ since it is impossible for both cats to be males unless one is. And, from the original sample space with four equally likely basic possible outcomes, one notes that $P(\text{mm}) = \frac{1}{4}$ and $P(\text{one cat is } m) = \frac{3}{4}$. So

$$P(\text{mm if one cat is m}) = \frac{P(\text{mm} \cap \text{“one cat is m”})}{P(\text{one cat is m})} = \frac{1/4}{3/4} = \frac{1}{3}.$$

(c) Similarly $P(\text{mm} \cap \text{“grey cat is m”}) = P(\text{mm}) = \frac{1}{4}$, and $P(\text{grey cat is m}) = \frac{1}{2}$. So

$$P(\text{mm if grey cat is m}) = \frac{P(\text{mm} \cap \text{“grey cat is m”})}{P(\text{grey cat is m})} = \frac{1/4}{1/2} = \frac{1}{2}.$$

The interesting feature of Examples 3 and 4 is that it makes a difference whether one knows that “one cat is male” or “the grey cat is male.” Look back again and see exactly why these two pieces of information are different.

Example 5. Statistics for the early 1980s indicate that 1 out of 3 adults in the United States smokes, that 1 out of 1500 adults will die of lung cancer within the next year, and that 1 out of 2000 both smoke and will die of lung cancer within the year. Are smoking and dying of lung cancer independent events? If not, compare the risks of lung cancer death for smokers and nonsmokers.

Solution. Let A be the event that an adult chosen at random smokes, and let B be the event that an adult chosen at random will die of lung cancer in the next year.

Then $P(A) = 1/3 = 0.33$, $P(B) = 1/1500 = 0.00067$, and $P(A \cap B) = 1/2000 = 0.0005$. So

$$P(A \cap B) \geq P(A) \cdot P(B),$$

which means that A and B are not independent events.

Furthermore,

$$P(\text{one dies of lung cancer in next year if one smokes})$$

$$= P(B \text{ if } A \text{ occurs}) = \frac{P(A \cap B)}{P(A)} = \frac{0.0005}{0.33} = 0.0015.$$

The work thus far has the following interpretation. With no extra information, you should assume that the probability of a person dying of lung cancer within a year is 0.00067. But, if you know that the person smokes, you should more than double this probability to 0.0015.

From the given data, you can also compute the probability of dying of lung cancer for a nonsmoker:

By the Corollary on page 185, $P(A^c \cap B) = P(B) - P(A \cap B)$. So

$$P(\text{one dies of lung cancer in the next year if one does not smoke})$$

$$= P(B \text{ if } A^c \text{ occurs}) = \frac{P(A^c \cap B)}{P(A^c)} = \frac{0.00067 - 0.0005}{1 - 0.33} = 0.00025.$$

Thus, if you know that a person smokes, then you appraise that person's risk of dying of lung cancer to be six times what it is for a nonsmoker.

The statistics used here, and in Problem 6 below, are generalized and simplified. They make no distinction between heavy smokers and occasional smokers, nor between former smokers and persons who have never smoked. And they do not consider the age, weight, sex, or life style of the subjects. But they do indicate the mathematical relationship between smoking and lung cancer for the general population.

When, as in Example 5, $P(B \text{ if } A \text{ occurs}) > P(B)$, one wants to conclude that the occurrence of event A "causes" or "enhances" the occurrence of event B. (In Example 5 this would be the conclusion that smoking causes lung cancer.) But mathematically this does not follow. In fact, one could just as well conclude that B causes A, as the following shows.

Theorem. *When $P(B \text{ if } A \text{ occurs}) > P(B)$, then $P(A \text{ if } B \text{ occurs}) > P(A)$.*

PROOF. The statement $P(B \text{ if } A \text{ occurs}) > P(B)$ means

$$\frac{P(A \cap B)}{P(A)} > P(B).$$

Thus $P(A \cap B) > P(A)P(B)$, and so

$$P(A \text{ if } B \text{ occurs}) = \frac{P(A \cap B)}{P(B)} > P(A). \qquad \square$$

Thus the concept of conditional probability carries no implications about cause and effect.

Regarding Example 5, all we have mathematically is a "correlation" between smoking and lung cancer. In Problem 6 below, you are asked to show that

$$P(\text{one smokes if one will die of lung cancer within the year}) = 0.75.$$

Thus, knowing nothing about an individual, you would say the probability that he or she smokes is 0.33. But, given the information that this person will die of lung cancer within a year, you increase your estimate of the probability that he or she is a smoker to 0.75.

A smoking advocate might argue that *perhaps* one's forthcoming death from lung cancer "causes" one to smoke. Or, *perhaps* some other factor altogether, affecting certain individuals, causes both lung cancer and addiction to nicotine.

But few people outside the tobacco industry would recommend that you stake your life on these alternative explanations.

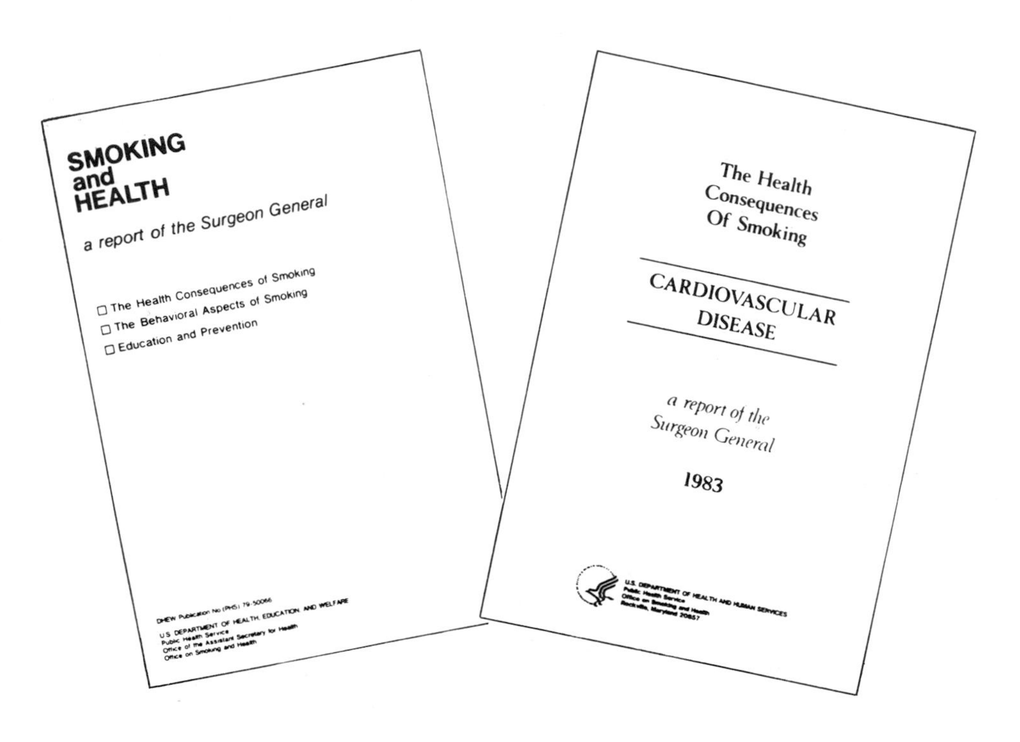

Problems

1. Two dice are rolled, and an observer announces that the two numbers are identical. Find the probability that the total score is (a) 4, (b) 7.

2. Two dice are rolled. What is the probability that the total score is 5, given that at least one die comes up 3.

3. A family has three children. Find the probability that all are boys if it is known that (a) the oldest is a boy, (b) at least one is a boy, (c) at least two are boys, (d) the two oldest are boys.

4. Let $P(A) = \frac{1}{4}$ and $P(A \cap B) = \frac{1}{6}$. If A and B are independent events, find $P(B)$.

5. Let $P(A) = \frac{1}{4}$ and $P(A \cap B) = \frac{1}{6}$. Does the occurrence of A make B more or less probable (a) if $P(B) = \frac{1}{2}$? (b) if $P(B) = \frac{3}{4}$?

6. Using the data of Example 5, compute P(one smokes if one dies of lung cancer in the next year), i.e., $P(A$ if B occurs).

7. Let $P(A) = \frac{1}{4}$, $P(A \cap B) = \frac{1}{6}$, and $P(A$ if B occurs$) = \frac{4}{5}$. Find (a) $P(B)$, (b) $P(B$ if A occurs), (c) $P(A$ if B does not occur).

8. What do you think of the following statement? In 1980 nearly half the people who died of heart disease were nonsmokers, so smoking probably does not cause heart disease. Explain your opinion.

9. One out of 300 adults in the United States died from heart disease in 1980, and about 53% of those were classified as smokers. Find
(a) P(one died of heart disease in 1980 if one smoked),
(b) P(one died of heart disease in 1980 if one did not smoke).

10. If you made your livelihood in the tobacco industry, how might you respond to the results of Problem 9?

*11. (J. Bertrand's Box Paradox, 1889). Three identical closed boxes are presented. One contains one silver coin and one gold coin, another contains two silver coins, and the third contains two gold coins. A person picks a box at random, reaches in blindfolded, and pulls out a coin. If the coin is gold, what is the probability that the remaining coin in the box is also gold?

[*Hint*. You could come up with answers of $\frac{2}{5}$, $\frac{1}{2}$, or $\frac{2}{3}$, depending on your point of view. To find out which (if any) of these is correct, carefully describe the sample space. Let the box with one silver and one gold coin be No. 1, and indicate its contents by s and g. Let the box with two silver coins be No. 2 and label its contents s_1 and s_2—two (distinguishable) silver coins. Let the box with two gold coins, g_1 and g_2, be No. 3. Then the equally likely basic possible outcomes can be represented by $1s, 1g, 2s_1, 2s_2, 3g_1$, and $3g_2$. For each basic outcome an integer 1, 2, or 3, identifying the box is followed by a letter identifying the coin removed.]

*12. If you solved Problem 11 with the aid of the definition of conditional probability, redo it without that definition. If you worked it without using the definition of conditional probability, redo it using that definition.

13. In order to learn whether fireflies of both sexes emit light, a small boy goes out at night and catches 20 fireflies in a suitable container (preserving their health). Later he observes that all are flashing. Does this prove (or suggest) that both sexes emit light?

14. Fifty five percent of fatal automobile accidents involve a driver within 25 miles of home. Does this suggest that an automobile trip near home is more dangerous than a trip across the country?

CHAPTER 13

Cardinality

This chapter is not easily "justified" in terms of day-to-day applications. It will not help you deal with compound interest, lay out the foundation for a house, decide which pizza is the better buy, appraise the risks of smoking, or judge the feasibility of civil defense plans.

The topic considered here—an examination of different "sizes of infinity"—has important applications in mathematics itself, and in the theory of computing (whether by machines or by human brains). But none of these applications will be explained here.

The reason for including a brief discussion of "cardinality" is that this offers a glimpse—at a fairly elementary level—of some of the questions and methods of reasoning which arise in "abstract" mathematics.

13.1. Countable Sets

Most people over the age of six have a pretty good idea how to count the number of elements in a given finite set. But it is not so easy to actually define what you mean by the words "one," "two," "three," ....

Try it!

And now the problem becomes even more difficult. This chapter considers the "number" of elements in some infinite sets.

Rather than trying to *define* the number of elements in a set, think first of merely *comparing* two sets to determine whether or not they have the same number of elements.

Definition. Two sets are said to have the **same cardinality** (number of elements) if they can be put into one-to-one correspondence with each other.

Example 1. The sets $A = \{a, b, c, d, e\}$ and $B = \{r, s, t, u, v\}$ have the same cardinality since there is the following one-to-one correspondence.

$$\begin{array}{cccccc} A: & a & b & c & d & e \\ & \updownarrow & \updownarrow & \updownarrow & \updownarrow & \updownarrow \\ B: & r & s & t & u & v \end{array}$$

Note that each element of set A is associated, via a double-headed arrow, with exactly one element of the set B and vice versa. There are many other ways to make a one-to-one correspondence between sets A and B. Simply reorder the elements of one of the sets.

In fact, the actual number of different one-to-one correspondences is $5! = 120$. (Confirm this.)

Now, one might define the *number of elements* in a nonempty finite set A to be that integer $n(A)$ such that the elements of A can be put into one-to-one correspondence with the elements of the set $\{1, 2, \ldots, n(A)\}$. This $n(A)$ is also called the *cardinality* of A.

For the sets in Example 1, $n(A) = n(B) = 5$.

The advantage of thinking in terms of one-to-one correspondences is that this concept provides a means of comparing infinite sets as well as finite ones.

Example 2. Show that the set of all nonnegative integers

$$A = \{0, 1, 2, 3, \ldots\}$$

has the same cardinality as the set of all positive integers

$$N = \{1, 2, 3, 4, \ldots\}.$$

Solution. One simply displays one of the many possible one-to-one correspondences:

$$\begin{array}{cccccc} N: & 1 & 2 & 3 & 4 & \ldots \\ & \updownarrow & \updownarrow & \updownarrow & \updownarrow & \\ A: & 0 & 1 & 2 & 3 & \ldots \end{array}$$

The symbol N will be used consistently to denote the set of all positive integers.

Definition. If the elements of a set A can be put into one-to-one correspondence with the positive integers, N, then the set A is said to be **countably infinite**. This is denoted symbolically by

$$n(A) = \aleph_0,$$

where $\aleph_0$ is pronounced "aleph nought."

Definition. If a set is either finite or countably infinite, it is said to be **countable**.

Example 3. Show that the set of *all* integers

$$Z = \{\ldots, -2, -1, 0, 1, 2, \ldots\}$$

is countable.

Solution. A little ingenuity is required to exhibit a one-to-one correspondence between the sets Z and N. If you attempted, say, to first use up the positive integers, in Z, you would never get to 0, or -1, or $-2, \ldots$. Instead, one must somehow alternate back and forth between positives and negatives.

Here is one way of doing it:

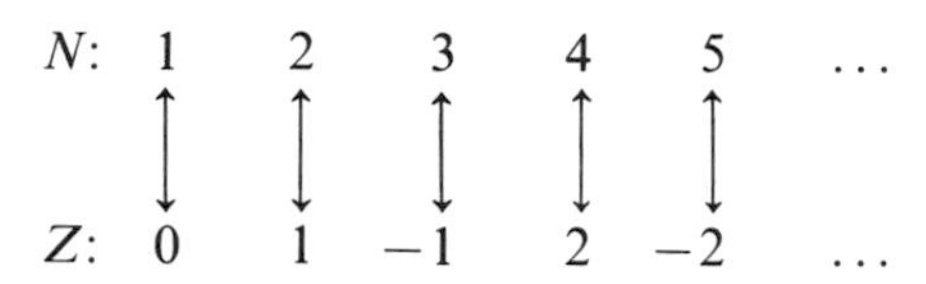

Looking at Examples 2 and 3, you might notice that we have really done more work than necessary. In order to demonstrate a one-to-one correspondence with the positive integers 1, 2, 3, ..., it would suffice to just write the second row in each case. That is, it is sufficient to just demonstrate how you could form an infinite "list" of the elements of the given set with a definite starting point—the "first" element.

So, in order to show that the set Z of all integers, considered in Example 3, is countable (or, more specifically, has cardinality $\aleph_0$), it would suffice to simply rewrite the elements of Z in the list

$$0, 1, -1, 2, -2, \ldots.$$

This shorter precedure will be used in the future.

Theorem A. *If A and B are countable sets, then $A \cup B$ is also countable.*

Proof. If either A or B is a finite set, then one can list all its elements before starting to list the elements of the other set.

If both A and B are countably infinite sets, say

$$A = \{a_1, a_2, a_3, \ldots\}$$

and

$$B = \{b_1, b_2, b_3, \ldots\},$$

then one can list the members of $A \cup B$ as

$$a_1, b_1, a_2, b_2, a_3, b_3, \ldots.$$

(In case A and B have any elements in common, the duplicates of any element already listed would simply be omitted when the same element is encountered again in the combined list.) □

Why, in the above proof, could one not avoid the use of subscripts by writing $A = \{a, b, c, \ldots\}$ and $B = \{p, q, r, \ldots\}$?

In a manner similar to the above proof, one can show that the union of any finite number of countable sets is again a countable set. See Problem 4 for the case of three sets.

PROBLEMS

1. Let $A = \{a, b, c, d\}$, $B = \{f, a, c, e\}$, and $C = \{d, c, g\}$. Find (a) $n(A \cup B)$, (b) $n(A \cap B)$, (c) $n(A \cup B \cup C)$, and (d) $n(A \cap B \cap C)$.
2. Exhibit another "list" of the elements of the set (a) A of Example 2, (b) Z of Example 3, and (c) $A \cup B$ in the proof of Theorem A.
3. Show that each of the following sets is countable.
   (a) $A = \{1/n: n = 1, 2, 3, \ldots\}$
   (b) $A = \{\ldots, -4, -2, 0, 2, 4, \ldots\}$
   (c) $A = \{\ldots, -4, -2, 0, 2, 4, \ldots\} \cup \{1, 3, 5, \ldots\}$
   (d) $A = \{1/n: n = 1, 2, 3, \ldots\} \cup \{2^n: n = 0, 1, 2, \ldots\}$.
4. Prove that if A, B, and C are three countable sets, then $A \cup B \cup C$ is also countable.
5. Show that N (the set of all positive integers) can be split into four disjoint sets, each of which has infinitely many members (elements).

13.2. Countably Many Countable Sets

Theorem A asserted that the union of a finite number of countable sets is again countable. The next result is stronger.

Theorem B. *The union of countably many countable sets is countable.*

PROOF. Let the given sets be $A_1, A_2, A_3, \ldots$, and denote the elements of these sets, using double subscripts, as follows:

$$A_1 = \{a_{11}, a_{12}, a_{13}, \ldots\}$$
$$A_2 = \{a_{21}, a_{22}, a_{23}, \ldots\}$$
$$A_3 = \{a_{31}, a_{32}, a_{33}, \ldots\}$$
$$\vdots$$

Thus a_{ij} is the jth element in set A_i.

(Why is it necessary to resort to double subscripts?)

Now the task is to form a single list of *all* elements of the union of these given sets.

One method for doing it is indicated in the following diagram. Relist the sets as given and then indicate, via arrows, an orderly scheme for creating a single list:

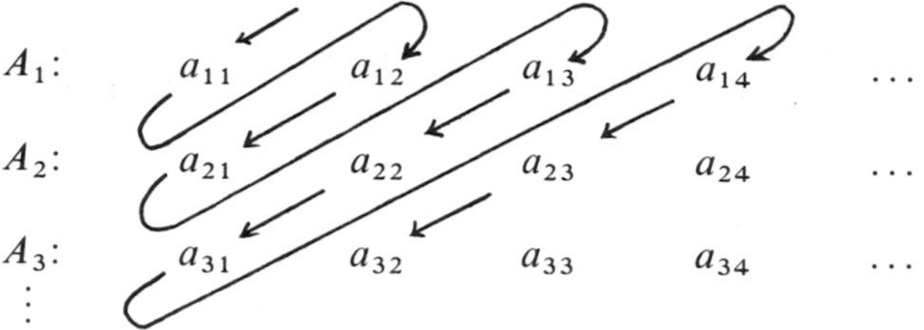

Following the arrows, one arrives at the new single list:

$$a_{11}, a_{12}, a_{21}, a_{13}, a_{22}, a_{31}, a_{14}, a_{23}, \ldots .$$

Once again, if any duplicates occur, they can be omitted. And if any of the sets $A_1, A_2, \ldots$ are finite, this merely "shortens" the final list. Thus the union of $A_1, A_2, \ldots$ is countable. □

Example 1. Show that the set of all *rational* numbers x such that $0 \le x \le 1$ is countable.

Solution. Here is a systematic scheme for listing these rationals (without duplicates):

$$0, 1, \tfrac{1}{2}, \tfrac{1}{3}, \tfrac{2}{3}, \tfrac{1}{4}, \tfrac{3}{4}, \tfrac{1}{5}, \tfrac{2}{5}, \tfrac{3}{5}, \tfrac{4}{5}, \tfrac{1}{6} \ldots .$$

Example 2. Show that the set of all positive rationals is countable.

Solution. Consider the set as the union of the following sets:

$$A_1 = \{1, 2, 3, \ldots\}$$
$$A_2 = \{\tfrac{1}{2}, \tfrac{3}{2}, \tfrac{5}{2}, \ldots\}$$
$$A_3 = \{\tfrac{1}{3}, \tfrac{2}{3}, \tfrac{4}{3}, \ldots\}$$
$$A_4 = \{\tfrac{1}{4}, \tfrac{3}{4}, \tfrac{5}{4}, \ldots\}$$
$$\vdots$$

Then invoke Theorem B.

Problems

1. Write the next five elements in the combined list started in the proof of Theorem B.
2. To prove Theorem B, why not begin by listing the first elements of each set, then the second elements of each set, etc.—as in the proof of Theorem A?
3. What would be the next two numbers in the list started in Example 1?
4. Prove that the set of all rationals (positive, negative, and zero) is countable. [You may use any previous examples and theorems.]

5. Prove that the set N (of all positive integers) can be split into infinitely many disjoint sets each of which has infinitely many elements.

13.3. The Reals Versus the Rationals

After reading the two previous sections, you might begin to suspect that every set is countable. The present section will show that this is not so.

A set which is not countable is said to be **uncountable**.

Notation. If a and b are numbers with $a < b$, the "intervals" $\{x: a < x < b\}$, $\{x: a \le x \le b\}$, and $\{x: a \le x < b\}$ are denoted by (a, b), $[a, b]$, and $[a, b)$, respectively.

[The symbol (a, b) was used earlier for an ordered pair of numbers. However, throughout this section (a, b) will mean an interval.]

Theorem C. *The set* $[0, 1)$ *is uncountable.*

Proof (due to Georg Cantor, 1845–1918). Note that it will *not* suffice to show that some particular list is incomplete. It must be shown that if *anyone* (no matter how talented) *ever* offers a list, claiming that it contains all the numbers in $[0, 1)$, that list must be incomplete.

Suppose (for contradiction) that there is a list of all the real numbers x such that $0 \le x < 1$. Let the list be $x_1, x_2, x_3, \ldots$. Each of these numbers can be expressed in decimal form. Denote the decimal representations as

$$x_1 = 0.a_{11}a_{12}a_{13}\ldots$$
$$x_2 = 0.a_{21}a_{22}a_{23}\ldots$$
$$x_3 = 0.a_{31}a_{32}a_{33}\ldots$$
$$\vdots$$

where each a_{ij} is a digit, 0, 1, 2, 3, 4, 5, 6, 7, 8, or 9.

A contradiction will be obtained by exhibiting some number in $[0, 1)$ which is missing from the list.

Such a number is

$$x = 0.b_1b_2b_3\ldots,$$

where

$$b_1 = \text{any digit } 1, 2, \ldots, \text{ or } 8 \text{ so long as } b_1 \ne a_{11},$$
$$b_2 = \text{any digit } 1, 2, \ldots, \text{ or } 8 \text{ so long as } b_2 \ne a_{22},$$
$$b_3 = \text{any digit } 1, 2, \ldots, \text{ or } 8 \text{ so long as } b_3 \ne a_{33},$$
$$\vdots$$

This construction pays particular attention to the digits "on the diagonal" in the proposed list, $x_1, x_2, \ldots$, namely, $a_{11}, a_{22}, a_{33}, \ldots$.

Now the number x is *not* one of the numbers in the list. For

$$\begin{aligned} x &\neq x_1 \quad \text{since} \quad b_1 \neq a_{11}, \\ x &\neq x_2 \quad \text{since} \quad b_2 \neq a_{22}, \\ x &\neq x_3 \quad \text{since} \quad b_3 \neq a_{33}, \\ &\vdots \end{aligned}$$

There is a technical detail in this proof which must be explained. Note that the choice of $b_1, b_2, b_3, \ldots$ always ruled out the digits 0 and 9. This is to avoid generating a number x which ends with repeating 0s or repeating 9s. Such numbers always have *another* correct decimal representation. For example, the number $0.25\overline{9}$ is equivalent to the fraction $234/900 = 13/50$. (Problem 6.) And that fraction in turn is equivalent to 0.26. So it follows that $0.25\overline{9}$ and $0.26\overline{0}$ are one and the same number, even though they do not look identical. One must avoid constructing a number x in the proof which might be identical to one of the listed numbers $x_1, x_2, x_3, \ldots$ even though it disagrees with each of these in some digit. □

Corollary. *The irrational numbers in* [0, 1), *or in* [0, 1], *are uncountable.*

PROOF. If these numbers were countable, then the set [0, 1) would be the union of two countable sets—its rational members and its irrational members—and would thus be countable itself. Theorem C says that this is not the case. □

Note that any subset of a countable set must also be countable. For, if one can "list" the elements of the original set, then one can proceed down this list, picking out for a new list those elements which belong to the subset. This yields the following.

Corollary. *The positive irrational numbers are uncountable and the set of all irrational numbers is uncountable.*

A Paradox. It is intuitively clear, and not particularly hard to prove, that between every two rational numbers there is an irrational, and between every two irrationals there is a rational. But, as has been shown, there are *far more* irrationals than rationals.

Example 1. Show that the set (0, 1) has the same cardinality as (−1, 1).

Solution. The following geometrical argument establishes a one-to-one correspondence between the two given sets. Draw parallel line segments representing (0, 1) and (−1, 1) as in Figure 1. Connect the left-hand end points and the right-hand end points and extend the resulting line segments until they intersect at a point P. Now draw rays through P which cut both line segments. The points of intersection, say $x \in (0, 1)$ and $y \in (-1, 1)$ are then a pair of points in the desired one-to-one correspondence.

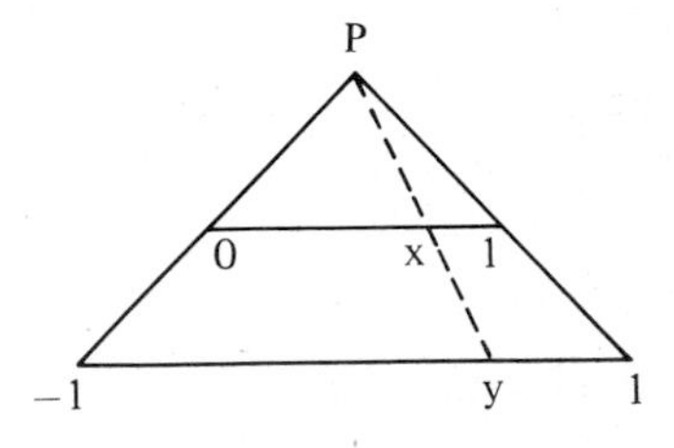

Figure 1.

You should be able to confirm, by an argument involving similar triangles, that $x/1 = (1 + y)/2$. So

$$y = 2x - 1 \qquad \text{for} \qquad 0 < x < 1.$$

This is the "analytical" expression of the one-to-one correspondence.

Example 2. Show that the set of all reals $(-\infty, \infty)$ has the same cardinality as $(-1, 1)$ and $(0, 1)$.

Solution. A one-to-one correspondence between the segments $[0, 1)$ and $[0, \infty)$ is exhibited in Figure 2. The segment $[0, 1)$ is now displayed vertically and $[0, \infty)$ is displayed horizontally. A broken line from the point P connects $x \in [0, 1)$ and $y \in [0, \infty)$ for a pair of corresponding points.

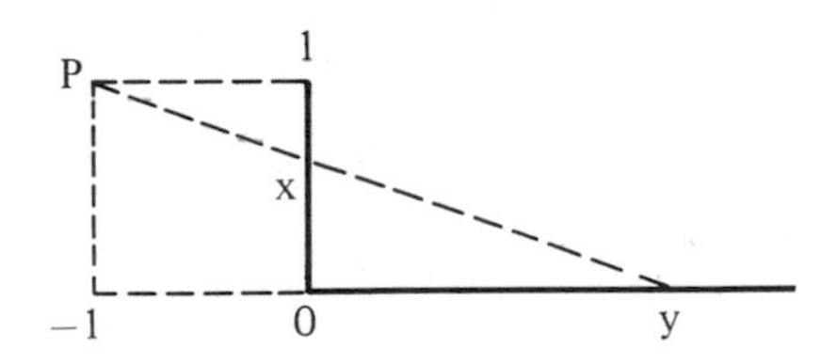

Figure 2.

By a similar construction you should be able to exhibit a one-to-one correspondence between $(-1, 0]$ and $(-\infty, 0]$. (Problem 4.)

These two constructions together establish the assertion that $(-\infty, \infty)$ has the same cardinality as $(-1, 1)$. And the fact that this has the same cardinality as $(0, 1)$ follows from Example 1.

The set $(-\infty, \infty)$, or any other set with the same cardinality, is said to have cardinality $\aleph_1$ (pronounced "aleph one"). One writes

$$n((-\infty, \infty)) = \aleph_1.$$

Thus $\aleph_1$ is another infinity—one which is "bigger than" $\aleph_0$.

It can be shown that there are always infinities "bigger than" any given infinity.

PROBLEMS

1. Consider the "list" of numbers in [0, 1] given in Example 1 of Section 13.2 (a) Is the number 1.4/2 in this list? (b) Is 1.414/2 in this list? (c) Is $\sqrt{2}/2$ in this list? (d) If not, why not? (e) Does this prove or disprove (or neither) that $n([0,1]) = \aleph_0$?
2. Explain your answer to Problem 1(e).
3. Identify three independent proofs, presented in this text, that not all numbers are rational.
4. Draw a figure, analogous to Figure 2, to exhibit a one-to-one correspondence between the sets $(-1,0]$ and $(-\infty,0]$.
5. Find the analytic expression for the one-to-one correspondence between [0, 1) and $[0,\infty)$ described geometrically in Figure 2.
6. Show that $0.25\overline{9} = 0.26$.
7. Show that the set (0, 1/2) is uncountable.
8. Exhibit both geometrically and analytically a one-to-one correspondence between (a) [1, 3] and [2, 7], (b) (1, 3) and (2, 7).

Answers to Odd-Numbered Problems

Section 1.1

1. (a) 14, (b) 12, (c) $3 + 3a$, (d) 55, (e) 130, (f) 1.7, (g) 106, (h) 0.006, (i) 39, (j) 17, (k) $a^2 - 2ab + b^2$, (l) $2x^2 - 13x - 7$ **3.** Only (b) is correct. **5.** (a) (1.05)10, (b) (1.15)39, (c) (0.9)27 **7.** $1 - (0.99)^2 = (1 + 0.99) \times (1 - 0.99) = (1.99)(0.01)$

Section 1.2

1. (a) $\frac{1}{3}$, (b) 3/37, (c) 17/10, (d) 14/3, (e) 2/125 **3.** (a) 5/2, (b) 2, (c) 3/20, (d) 5, (e) 1/2, (f) 9/4 **5.** (a) 4/9, (b) 7/8, (c) 15/19 **7.** Only (d) is correct **9.** the 25-ounce package **11.** Yes, if the commodity in question is one that you use regularly and you have only one (or a small number) of these coupons. **13.** (a) 128, (b) 81, (c) 1/32, (d) 100, (e) 2, (f) 256 **15.** Each is wrong. **17.** 2.54

Section 1.3

1. $1360 **3.** (a) 1.07, (b) 1.08, (c) 1.055 **5.** (a) 1.026, a 2.6% increase, (b) 0.963, a 3.7% decrease **7.** 150% **9.** 23% (approximately) **11.** (a) 20% increase, (b) $66\frac{2}{3}$% increase, (c) 17.5% decrease **13.** 20% **15.** $33\frac{1}{3}$% **17.** 5%

Section 1.4

1. (a) 36 miles, (b) 132 feet **3.** (a) 1/30, (b) 0.27 cents **5.** $11.20. For greatest savings, turn off the heater *before* the last shower or two to use up the water you have already heated. **7.** 52 gallons **9.** (a) 36 minutes, (b) 4.1 seconds **11.** About $\frac{1}{2}$ second **13.** (a) 17.2, (b) 0.058

Section 2.1

1. 2, 3, 5, 7, 11, 13, 17, 19, 23, 29 **3.** (a) $3^4 \times 23$, (b) $2^3 \times 3 \times 13 \times 17$, (c) $5^4 \times 11$, (d) prime, (e) 11×17, (f) $3 \times 5 \times 499$ (Show that 499 is prime.) **5.** when the last three digits form a number divisible by 8 **7.** (a) 3277/9776, (b) $-73/7560$ **9.** (a) and (d) are divisible by 11, (b) and (c) are not.

Section 2.2

1. (a) 47/103, (b) 119/59, (c) 73/103 **3.** (a) more than 10^{35} years, (b) more than 10^{33} years (So encryptions based on products of large primes should be quite secure.)

Section 2.3

1. (a) 0.375, (b) $0.\overline{1}$, (c) $1.\overline{18}$, (d) $0.\overline{571428}$, (e) $1.\overline{216}$ **3.** (a) 9/11, (b) 10/9, (c) 1326/55, (d) 31/37, (e) 1/7, (f) 1 **5.** b-1 digits

Section 3.1

1. 13 **3.** 25 feet **5.** 240 feet **7.** acute **9.** (a) 110, (b) 1.1, (c) 800, (d) 0.08, (e) 8/9

Section 3.2

1. $\sqrt{3}$ **3.** Suppose (for contradiction) that $\sqrt{3}$ is rational. Then $\sqrt{3} = m/n$ where m and n are integers not both divisible by 3. Find $m^2 = 3n^2$, so that m^2 is divisible by 3. Then, by Problem 2, $m = 3p$ for some integer p. This gives $9p^2 = 3n^2$; and it follows that n^2 and hence also n are divisible by 3. This gives the desired contradiction. **5.** You will be unable to prove the analogue of the lemma. Why?

Section 3.3

1. 1.4142 **3.** 2.236 **5.** (a) 14.14, (b) 0.1732, (c) 223.6, (d) 0.577 **7.** 23 feet **9.** 0.5%

Section 4.1

1. 16 and 21 years **3.** $50,000 **5.** 13 ounces **7.** $2\frac{1}{2}$ liters **9.** 3 pints **11.** 75 km/hour **13.** 1 hour 20 minutes **15.** 60 feet **17.** 3 feet 9 inches from the pivot (on the other side) **19.** The lever will eventually bend or break, or the fulcrum will sink into the ground. **21.** 30 and 10 years

Section 4.2

1. 16 and 21 years **3.** 9 and 14 years **5.** 6.25 ounces at 2% and 3.75 ounces at 10% **7.** 7.5 mph and 2.5 mph

Section 4.3

1. See figure below. **3.** (a) (2/3, 11/6), (b) (-14, $-27/5$), (c) no solution, (d) infinitely many solutions **5.** (a) $y = 0.5x - 10{,}527$, (b) See figure below. (c) \$39,473 **7.** No. Explain.

1.

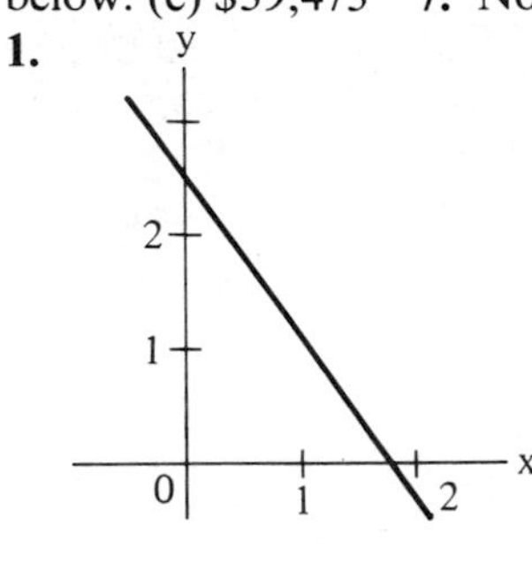

Sec. 4.3 #1.

5.(b)

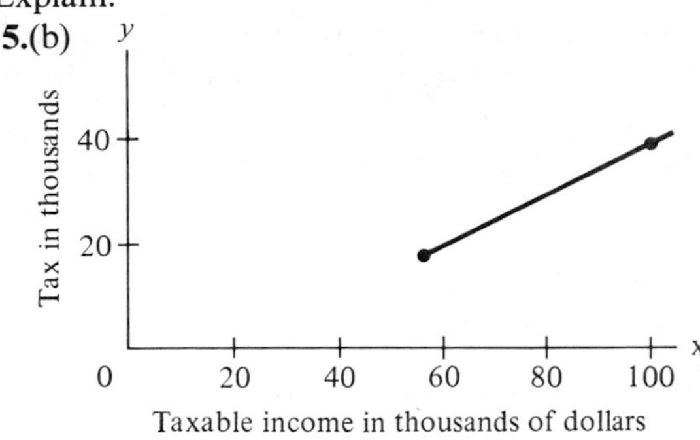

Sec. 4.3 #5b.

Section 5.1

1. (a) $-4, 1$, (b) 5 (only), (c) $\pm\sqrt{7}$, (d) $2, -\frac{1}{2}$, (e) $\frac{5}{2}, -\frac{1}{3}$, (f) $-1 \pm \sqrt{3}$, (g) no real solutions

Solution 5.2

1. 16 and 17, or -17 and -16 **3.** 150 feet by 200 feet **5.** 25 mph **7.** 18 **9.** (a) after $(5 \pm \sqrt{5})/2$ seconds, (b) after 2.5 seconds, (c) never **11.** never

Section 5.3

1. $y = 1/4$ and $1/16$ **3.** (a) $x = -1 \pm \sqrt{10}/2$, (b) $x = (5 \pm \sqrt{57})/4$ **5.** (a) 40 feet, (b) after $(3 + \sqrt{10})/2$ seconds **7.** 150 feet after 2.5 seconds

Section 6.1

1. (c) 2.65 miles **3.** (a) about 4, (b) 5, (c) 7 **5.** (a) 90°F, (b) after yet another 3 days, (c) never **7.** (d) This will be discussed further in Section 12.5.

Section 6.2

1. (a) $r\sqrt{r}$ and $r^2\sqrt{r}$, (b) 0.35 and 0.18 **3.** about 2300 B.C. **5.** about 96°F **7.** about 13,600 years **9.** Let the store get warmer. It takes much more energy to cool the entire store than just to cool the freezers to a given temperature.

Section 6.3

1. (a) \$8.00, (b) \$8.95, (c) \$9.07 **3.** (a) \$3000, (b) \$6727.50, (c) \$7209.57 **5.** about 1 billion **7.** (a) about the year 2070, (b) about eight billion **9.** 9.2% **11.** (a) 10.38%, (b) 10.47%, (c) 10.52%, (d) 10.67%.

Section 6.4

Answers rounded to nearest dollar for readability: **1.** (a) \$22,178,

(b) \$11,529 **3.** (a) \$2109, (b) about \$1460 **5.** (a) \$500, (b) \$520 **7.** (a) \$498, (b) \$457 **9.** Let the effective annual interest rates divided by 100 be i_1 the first year, i_2 the second year, and so on. Then after n years an IRA yields $\$2000(1 + i_1)(1 + i_2) \cdots (1 + i_n)(1 - t)$ after taxes. An ordinary account yields only $\$2000(1 - t)[1 + i_1(1 - t)][1 + i_2(1 - t)] \cdots [1 + i_n(1 - t)]$.

Section 6.5

1. 8 **3.** 410/333 **5.** The annual salary would be \$45 trillion—many times the national debt. **7.** (a) 20/3 feet, (b) $\frac{3}{2}$ seconds **9.** $\frac{1+r}{1-r}$h

Section 7.1

1. (a) 144, (b) 1296, (c) 9, (d) 10^4, (e) 10^6, (f) 6.45 **3.** \$270 **5.** 2400 **7.** (a) $220\sqrt{3}$ by $198\sqrt{3}$ feet, (b) $\frac{1}{4}$ acre, (c) 640 **9.** 14 inches **11.** 56.25% more **13.** 6 gallons

Section 7.2

1. (a) 27, (b) 46,656, (c) 10^6, (d) 16.4 **3.** 9.3 ounces **5.** $5\frac{1}{3}$ **7.** 37 (you cannot order 36.74 "yards.") **9.** approximate answers: (a) 1000, (b) 10,000, (c) 125 tons **11.** 0.7 inch **13.** about $3\frac{1}{2}$ teaspoons

Section 7.3

1. $A = 2(ab + ac + bc)$ **3.** 220 **5.** (a) 25%, (b) 25% **7.** \$250.00 **9.** Think of the person as an ant enlarged by a multiplier, say $m = 300$. Then the person's weight will be 27 million times that of the ant, while his or her surface area (which determines air resistance) will be only 90 thousand times that of the ant. **11.** (a) 32 billion acres, (b) about 7 acres, (c) less than 175 square feet

Section 7.4

1. 1.26 **3.** 1.59 **5.** about 5 ounces **7.** about 3.8 inches **9.** (a) 6.7 inches, (b) 9.7 inches **11.** 12.25% **13.** (a) 2.52, (b) 3.17, (c) 0.85, (d) 0.48, (e) 1.82, (f) 2.15 **15.** 17

Section 8.1

1. $\sqrt{5} \cong 2.24$ miles **3.** 98 feet/second and 78 feet/second **5.** (a) 170 mph, (b) 70 mph, (c) 130 mph, (d) 109.1 mph **7.** (a) 10 seconds, (b) 12.5 seconds **9.** $(v^2 - w^2)/v$ mph **11.** white. But few people (including mathematicians) can give a correct explanation if they have not already seen it somewhere.

Section 8.2

1. about 2 miles **3.** 5.9×10^{12} miles **5.** (a) 300 meters, (b) 3 meters **7.** approximately 216 and 184 cycles/second **9.** Frequency is 3×10^{-5}% *lower* than sent.

Section 8.3

1. 195 miles **3.** 70 mph **5.** Too low; only the component of the target vehicle's velocity toward the observer matters, and this is smaller than the vehicle's actual speed. See Figure 8. **7.** (a) a continuous Doppler decrease in pitch, (b) no change or possible increase in pitch

Section 9.1

1. $t - 2d/c$ **3.** 6800 **5.** (a) no, (b) 16×10^9 years **7.** (a) 5×10^{-6} seconds, (b) 5.000001×10^{-6} seconds

Section 9.2

1. about 8 years and 8 months **3.** 0.995c

Section 9.3

1. $v = \sqrt{3}c/2$ **3.** (a) 3.125×10^{-8} seconds, (b) 18.4 feet, (c) It finds the distance to be only 14.7 feet.

Section 10.1

1. (a) 10011 and 101010, (b) 201 and 1120 **3.** (a) 10100, (b) 100011 **5.** (a) 122101, (b) 12222 **7.** (a) 14, (b) 9

Section 10.2

1. (a) 101, (b) 10 **3.** (a) $\frac{5}{4}$ or 1.25, (b) 19/8 or 2.375 **5.** (a) 1/11, (b) 111/101, (c) 1

Section 10.3

1. 256 **3.** (a) 2, (b) 6, (c) 14, (d) 30, (e) 62 **5.** (a) FADE, (b) FEED, (c) BEAD or DEAD

Section 11.1

1. Many answers are possible. For example $A = \{4, 2, 6\}$, $A = \{x\text{: } x \text{ an even integer}, 0 < x \leq 7\}$, $B = \{1, 2\}$, $B = \{2, 1\}$, $C = \{x\text{: } x \text{ a prime number} < 12\}$, $C = \{5, 2, 11, 7, 3\}$ **3.** $S = \{$(h, h, h), (h, h, t), (h, t, h), (h, t, t), (t, h, h), (t, h, t), (t, t, h), (t, t, t)$\}$, where the first, second, and third letters in each ordered triple represent the outcomes of the first, second, and third tosses, respectively. **5.** $\{-2\}$

Section 11.2

1. 16 **3.** 216 **5.** (a) 22,100, (b) 4

Section 12.1

1. (a) 1/13, (b) $\frac{1}{4}$, (c) 4/13 **3.** (a) 8, (b) {hht, hth, thh}, (c) 3/8 **5.** three of one sex and one of the other **7.** If the elevator goes up and down between the first and twelfth floors at a constant rate, then

$$P(\text{elevator is going down as it arrives at ninth floor}) = 3/11.$$

Section 12.2

1. 0.9992 **3.** (a) 5/18, (b) 13/18 **5.** (a) 1/36, (b) 25/36, (c) 11/36, (d) 5/18 **7.** 1/221 **9.** (a) 1,712,304, (b) 0.66, (c) 0.34 **11.** 1/12

Section 12.3

1. (a) 0.669, (b) 0.331 **3.** (a) 7/8, (b) 15/16 **5.** (a) 0.96, (b) 0.32 **7.** 0.518 **9.** (a) 0.635, (b) 0.456, (c) 0.821 **11.** (a) 0.865, (b) 0.9997 (This makes no allowance for driver errors or collisions.)

Section 12.4

1. (a) 0.69, (b) 8 **3.** (a) about 0.34, (b) about 0.66 **5.** (a) about 0.12, (b) about 0.27 **7.** about 0.91

Section 12.5

1. $\frac{1}{2}$ **3.** (a) 3/20, (b) 15/4 **5.** (a) 0.04, (b) 0.32, (c) 0.64, (d) 1.6 **7.** \$92.00 **9.** (a) If x is your gain in dollars, then $x = 99{,}999$, or $x = 9{,}999$, or $x = -1$. Clearly $P(x{=}99{,}999) = 1/150{,}000$. If you win the first prize, you will not also be considered for the second prize, so $P(x{=}9{,}999) = (149{,}999/150{,}000)(1/149{,}999) = 1/150{,}000$. It now follows that $P(x{=}-1) = 1 - 2/150{,}000$. So $E(x) = 99{,}999/150{,}000 + 9{,}999/150{,}000 - 149{,}998/150{,}000 = -0.27$. (b) 74,999 to 1 **11.** $P(x{=}237.5) = \frac{1}{8}$, $P(x{=}12.5) = \frac{3}{8}$, $P(x{=}-62.5) = \frac{3}{8}$, $P(x{=}-87.5) = \frac{1}{8}$, $E(x) = 0$.

Section 12.6

1. (a) $\frac{1}{6}$, (b) 0 **3.** (a) $\frac{1}{4}$, (b) $\frac{1}{7}$, (c) $\frac{1}{4}$, (d) $\frac{1}{2}$ **5.** (a) more, (b) less **7.** (a) 5/24, (b) 2/3, (c) 2/19 **9.** (a) 0.0053, (b) 0.0024 **11.** 2/3 **13.** not in the least. Explain.

Section 13.1

1. (a) 6, (b) 2, (c) 7, (d) 1 **3.** (a) trivial, (b) much like Example 3, (c) and (d) invoke Theorem A. **5.** $A = \{1, 5, 9, \ldots\}$, $B = \{2, 6, 10, \ldots\}$, $C = \{3, 7, 11, \ldots\}$, $D = \{4, 8, 12, \ldots\}$

Section 13.2

1. $a_{32}, a_{41}, a_{15}, a_{24}, a_{33}$ **3.** $\frac{5}{6}, \frac{1}{7}$

5.

$A_1 = \{1, 2, 4, 7, 11, \ldots\}$

$A_2 = \{3, 5, 8, 12, \ldots\}$

$A_3 = \{6, 9, 13, \ldots\}$

A_4 & $\{10, 14, \ldots\}$

$\vdots$

{The arrows give a clue for continuing the construction.

Section 13.3

1. (a) yes, (b) yes, (c) no, (d) The list includes only rationals, (e) neither **3.** Section 2.3 gave examples of irrationals, Section 3.2 gave a proof that $\sqrt{2}$ is irrational, and the present chapter has shown that the rationals are countable while the reals are not. **5.** $y = x/(1 - x)$ for $0 \leq x < 1$ **7.** If it were countable, then by an argument like that in Example 1, you could show that $(-1, 1)$ and hence its subset $[0, 1)$ would be countable, contradicting Theorem C.

Index

[*Note*. The page listed may be followed by others on the topic. Some items in the list of Applications on pages ix–xii are not repeated here.]

Airplane speeds
  across wind, 127, 141
  with or against wind, 49, 54, 70, 127
Animal sizes, 117
Archimedes, 112
Area, 103
  acre, 107
  circle, 106
  cylinder, 115
  definition, 104
  enclosed by fence, 74
  rectangle, 34, 103
  similar regions, 105, 115
  sphere, 117
  surface, 115
  triangle, 34, 106
Associative laws, 2
Astronomy
  age of universe, 147
  distances, 134, 147
  red shift, 134
Automobile
  antifreeze, 46
  breakdowns, 193
  fatalities, 211
  gas mileage, 20
  speed and distance, 45

Ball bouncing, 99
Bear, color of, 127
Bertrand's box paradox, 211
Binary numbers, 157
Bit, 166
Boat speeds
  across river, 124, 137
  up and down river, 54, 67, 127
Bomb falling, 141
Byte, 166

Cantor, Georg, 217
Carbon dating, 82
Cardinality, 212
  aleph nought, 213
  aleph one, 219
  countable sets, 214
  rationals, 216
  reals, 217
  uncountable sets, 217
Celsius-Fahrenheit, 60
Chain letter, 78, 99
Civil defense
  crisis relocation, 193
  fallout shelter, 114
Clocks, 143, 148

Coin
sizes, 112, 116, 122
tossing, 171, 178, 206
Color, 134
Commutative laws, 1
Completing the square, 63
Compound interest, 87, 92, 102
Consumer Price Index, 90
Cooling, Newton's law, 79, 84
Corners, square, 35
Counting, 174, 212
combinations, 176
permutations, 175
Coupons, 11
Cube roots, 118
Currency conversion, 45

Data
encryption, 29
transmission, 134, 147, 164
Difference of two squares, 5
Discounts, 13
Distributive law, 3
Divisibility, 22
Division algorithm, 27
Doppler effect, 128, 138
bomb falling, 141
observer moving, 129
radar speed trap, 132
source moving, 130
Down payments, 12, 48

Einstein, A., 143, 145
Electric consumption, 16
Elementary particles, 150, 154
Elevator arrival, 182
Energy
conservation, 19, 87
heat losses, 79, 86, 118
solar heat storage, 80, 86, 117
Equations
linear in one unknown, 44
quadratic, 62
simultaneous linear, 50
Ether, 145
Euclid, 24, 28, 34
Exponents, 9

Factors
common, 23, 26
of polynomials, 63, 76
prime, 21
Faucet leak, 19
Fermat, P., 193
Fireflies, 211
Fitzgerald, G. F., 153
Flower pots, 114, 120
Fractions, 6, 23, 27, 30, 99
Free fall, 69, 75, 100
Frequency and wavelength, 128
color of light, 134
pitch of sound, 129
radio, 134

Galilean relativity, 123
Gambling, 81, 193, 202
coins, 180
dice, 183
fair bet, 203
fair game, 203
lottery, 205
poker hands, 176, 186
odds, 203
ruin, 204
Garfield, J. A., 34
Geometric sequence, 78
Geometric series, 96, 163
Gold testing, 112
Graphs, 54, 71, 83
Gravity, 69, 101
Greatest common factor, 23, 26

Half lives, 78, 81
Health and smoking, 185, 208
Heart disease, 211
Heat loss, 80, 86, 116
Hieron II, 112
Hoses
flow rates, 107
material, 117

Individual Retirement Accounts, 91
Infinities, 213, 219
Interest, 18

compound, 87
deferred, 89, 120
rates, 18
statements, 18
tax free, 93

Kilowatt-hour, 17

Lawn, 107, 114
Least common multiple, 24
Length contraction, 151
Levers, 47
lifting, 48
scales, 49
seesaw, 49
Libby, W., 82
Life insurance, 201
Light
color, 134
red shift, 134
speed, 132
wavelength, 134
Linear relation, 56
Lorentz, H. A., 143, 153
Lumber sizes, 42, 113
Lung cancer, 185, 208

Materials
black dirt, 114
can, 117, 121
carpet, 107
concrete, 114
drapes, 107
fertilizer, 107, 114
grass seed, 107
gravel, 113
hose, 117
lumber, 42, 113
paint, 108
pizza, 106
string, 111
yarn, 107, 113, 121
Mesons, 150, 154, 155
Metric units
Celsius, 60
centimeter, 11, 107
kilogram, 114
liter, 19
meter, 11
Mixtures, 46, 52
Monte Carlo fallacy, 193, 206
Morse code, 165
Mortality statistics, 186, 201
Musical tones, 129, 131, 134

Newton's law of cooling, 79
n'th root computation, 122
Numbers
base two, 157
binary, 157
irrational, 32, 38, 218
prime, 21
rational, 32, 216
ternary, 157

Parity bit, 167
Pascal, B., 193
Percentages, 11
Poincare, H., 153
Polynomials, 71, 76
Population growth, 89, 91, 118
Powers, 9, 77
large, 88, 90, 193
of ½, 83
Prime numbers, 21
Probability, 178
automobile breakdowns, 193
cards, 179, 185, 187
coins, 178, 187, 206
conditional, 207
dice, 178, 183, 193
elevator arrival, 182
event, 180
expectation, 200
fireflies, 211
independent events, 207
life insurance, 201
mutually exclusive events, 183
odds, 203
poker hands, 175, 186
quality control, 192, 194
rules, 181, 183, 188
sample space, 179

Probability (*cont.*)
sex of children, 182, 206
smoking and health, 185, 208
Pyramid
clubs, 80, 102
Great, 37
Pythagorean Theorem, 34
converse, 35

Quadratic
equation, 62
formula, 65
polynomial, 71
Quality control, 192, 194

Radar speed trap, 132, 141
Radio waves, 132
frequencies, 134
speed, 132
Radioactive materials
decay, 78, 82
waste, 86
Rates
dividend, 51
drip, 19
electric, 16
interest, 18
miles per gallon, 20
speed, 16
water flow, 107
Red shift, 134
Relativity
Galilean, 123
length contraction, 151
special, 143
time dilation, 148
Repeating decimal, 31, 98

Sailing, 139
Sales charges, 14, 91
Sales tax, 12, 48
Sequence, geometric, 78
Series, geometric, 96, 163
Sets, 170
countable, 214
counting, 174
uncountable, 217
Simultaneity, 143
Sizes
balls of yarn, 111, 113
coins, 112, 116, 122
containers, 111, 117, 120
hoses, 107, 113
lumber, 42, 113
pizzas, 106
Smoking and health, 185, 208
Solar heat storage, 80, 86, 117
Sound
Doppler effect, 128, 138
pitch, 129
speed, 20, 49
wavelength, 128
Space travel
relativity, 150, 153
Voyager, 134, 147, 167
Special relativity, 143
Speed, 16, 49, 53, 67, 123
airplane, 49, 54, 70, 127, 141
average, 45, 127
boat, 54, 67, 124, 137
light, 132, 145
sound, 20, 49, 128
Square corners, 35, 43
Square roots, 36
computation, 41
irrational, 38
Squares, difference of, 5
Successive approximations, 41, 119, 122

Tax
income, 15, 60, 91
sales, 12, 48
shelter, 91
Television sizes, 35, 49, 122
Temperature scales, 60
Time, 144, 148

Vectors
addition, 124
components, 135
displacement, 124, 135

force, 139
velocity, 125, 137
Volume, 108
cylinder, 111
definition, 109
irregular, 112
rectangular, 108
similar solids, 110
sphere, 111

Wavelength, 128

Zero coupon CD, 89, 120

Undergraduate Texts in Mathematics

continued from ii